SCIENCE 200

Teacher's Guide

Author:
Alpha Omega Publications

Editor:
Alan Christopherson, M.S.

804 N. 2nd Ave. E.
Rock Rapids, IA 51246-1759

SCIENCE 200

LIFEPAC® Overview

SCIENCE SCOPE & SEQUENCE

	Your Senses (Grade 1)	Our World (Grade 2)	Changes to Our World (Grade 3)
Unit 1	YOU LEARN WITH YOUR EYES • Name and group some colors • Name and group some shapes • Name and group some sizes • Help from what you see	THE LIVING AND NONLIVING • What God created • Rock and seed experiment • God-made objects • Man-made objects	YOU GROW AND CHANGE • Air we breathe • Food for the body • Exercise and rest • You are different
Unit 2	YOU LEARN WITH YOUR EARS • Sounds of nature and people • How sound moves • Sound with your voice • You make music	PLANTS • How are plants alike • Habitats of plants • Growth of plants • What plants need	PLANTS • Plant parts • Plant growth • Seeds and bulbs • Stems and roots
Unit 3	MORE ABOUT YOUR SENSES • Sense of smell • Sense of taste • Sense of touch • Learning with my senses	ANIMALS • How are animals alike • How are animals different • What animals need • Noah and the ark	ANIMAL AND ENVIRONMENT CHANGES • What changes an environment • How animals are different • How animals grow • How animals change
Unit 4	ANIMALS • What animals eat • Animals for food • Animals for work • Pets to care for	YOU • How are people alike • How are you different • Your family • Your health	YOU ARE WHAT YOU EAT • Food helps your body • Junk foods • Food groups • Good health habits
Unit 5	PLANTS • Big and small plants • Special plants • Plants for food • House plants	PET AND PLANT CARE • Learning about pets • Caring for pets • Learning about plants • Caring for plants	PROPERTIES OF MATTER • Robert Boyle • States of matter • Physical changes • Chemical changes
Unit 6	GROWING UP HEALTHY • How plants and animals grow • How your body grows • Eating and sleeping • Exercising	YOUR FIVE SENSES • Your eye • You can smell and hear • Your taste • You can feel	SOUNDS AND YOU • Making sounds • Different sounds • How sounds move • How sounds are heard
Unit 7	GOD'S BEAUTIFUL WORLD • Types of land • Water places • The weather • Seasons	PHYSICAL PROPERTIES • Colors • Shapes • Sizes • How things feel	TIMES AND SEASONS • The earth rotates • The earth revolves • Time changes • Seasons change
Unit 8	ALL ABOUT ENERGY • God gives energy • We use energy • Ways to make energy • Ways to save energy	OUR NEIGHBORHOOD • Things not living • Things living • Harm to our world • Caring for our world	ROCKS AND THEIR CHANGES • Forming rocks • Changing rocks • Rocks for buildings • Rock collecting
Unit 9	MACHINES AROUND YOU • Simple levers • Simple wheels • Inclined planes • Using machines	CHANGES IN OUR WORLD • Seasons • Change in plants • God's love never changes • God's Word never changes	HEAT ENERGY • Sources of heat • Heat energy • Moving heat • Benefits and problems of heat
Unit 10	WONDERFUL WORLD OF SCIENCE • Using your senses • Using your mind • You love yourself • You love the world	LOOKING AT OUR WORLD • Living things • Nonliving things • Caring for our world • Caring for ourselves	PHYSICAL CHANGES • Change in man • Change in plants • Matter and time • Sound and energy

SCIENCE SCOPE & SEQUENCE

God's Creation (Grade 4)	Life Cycles and Geology (Grade 5)	Plants and Animals/Space (Grade 6)	
PLANTS • Plants and living things • Using plants • Parts of plants • The function of plants	CELLS • Cell composition • Plant and animal cells • Life of cells • Growth of cells	PLANT SYSTEMS • Parts of a plant • Systems of photosynthesis • Transport systems • Regulatory systems	Unit 1
ANIMALS • Animal structures • Animal behavior • Animal instincts • Man protects animals	PLANTS: LIFE CYCLES • Seed producing plants • Spore producing plants • One-celled plants • Classifying plants	BODY SYSTEMS • Digestive system • Excretory system • Skeletal system • Diseases	Unit 2
MAN'S ENVIRONMENT • Resources • Balance in nature • Communities • Conservation and preservation	ANIMALS: LIFE CYCLES • Invertebrates • Vertebrates • Classifying animals • Relating function and structure	PLANT AND ANIMAL BEHAVIOR • Animal behavior • Plant behavior • Plant and animal interaction • Balance in nature	Unit 3
MACHINES • Work and energy • Simple machines • Simple machines together • Complex machines	BALANCE IN NATURE • Needs of life • Dependence on others • Prairie life • Stewardship of nature	MOLECULAR GENETICS • Reproduction • Inheritance • DNA and mutations • Mendel's work	Unit 4
ELECTRICITY AND MAGNETISM • Electric current • Electric circuits • Magnetic materials • Electricity and magnets	TRANSFORMATION OF ENERGY • Work and energy • Heat energy • Chemical energy • Energy sources	CHEMICAL STRUCTURE • Nature of matter • Periodic Table • Diagrams of atoms • Acids and bases	Unit 5
PROPERTIES OF MATTER • Properties of water • Properties of matter • Molecules and atoms • Elements	RECORDS IN ROCK: THE FLOOD • The Biblical account • Before the Flood • The Flood • After the Flood	LIGHT AND SOUND • Sound waves • Light waves • The visible spectrum • Colors	Unit 6
WEATHER • Causes of weather • Forces of weather • Observing weather • Weather instruments	RECORDS IN ROCK: FOSSILS • Fossil types • Fossil location • Identifying fossils • Reading fossils	MOTION AND ITS MEASUREMENT • Definition of force • Rate of doing work • Laws of motion • Change in motion	Unit 7
THE SOLAR SYSTEM • Our solar system • The big universe • Sun and planets • Stars and space	RECORDS IN ROCK: GEOLOGY • Features of the earth • Rock of the earth • Forces of the earth • Changes in the earth	SPACESHIP EARTH • Shape of the earth • Rotation and revolution • Eclipses • The solar system	Unit 8
THE PLANET EARTH • The atmosphere • The hydrosphere • The lithosphere • Rotation and revolution	CYCLES IN NATURE • Properties of matter • Changes in matter • Natural cycles • God's order	ASTRONOMY AND THE STARS • History of astronomy • Investigating stars • Major stars • Constellations	Unit 9
GOD'S CREATION • Earth and solar system • Matter and weather • Using nature • Conservation	LOOK AHEAD • Plant and animal life • Balance in nature • Biblical records • Records of rock	THE EARTH AND THE UNIVERSE • Plant systems • Animal systems • Physics and chemistry • The earth and stars	Unit 10

SCIENCE SCOPE & SEQUENCE

	General Science I (Grade 7)	General Science II (Grade 8)	General Science III (Grade 9)
Unit 1	WHAT IS SCIENCE? • Tools of a scientist • Methods of a scientist • Work of a scientist • Careers in science	SCIENCE AND SOCIETY • Definition of science • History of science • Science today • Science tomorrow	OUR ATOMIC WORLD • Structure of matter • Radioactivity • Atomic nuclei • Nuclear energy
Unit 2	PERCEIVING THINGS • History of the metric system • Metric units • Advantages of the metric system • Graphing data	STRUCTURE OF MATTER 1 • Properties of matter • Chemical properties of matter • Atoms and molecules • Elements, compounds, and mixtures	VOLUME, MASS, AND DENSITY • Measure of matter • Volume • Mass • Density
Unit 3	EARTH IN SPACE 1 • Ancient stargazing • Geocentric Theory • Copernicus • Tools of astronomy	STRUCTURE OF MATTER 2 • Changes in matter • Acids • Bases • Salts	PHYSICAL GEOLOGY • Earth structures • Weathering and erosion • Sedimentation • Earth movements
Unit 4	EARTH IN SPACE 2 • Solar energy • Planets of the sun • The moon • Eclipses	HEALTH AND NUTRITION • Foods and digestion • Diet • Nutritional diseases • Hygiene	HISTORICAL GEOLOGY • Sedimentary rock • Fossils • Crustal changes • Measuring time
Unit 5	THE ATMOSPHERE • Layers of the atmosphere • Solar effects • Natural cycles • Protecting the atmosphere	ENERGY 1 • Kinetic and potential energy • Other forms of energy • Energy conversions • Entropy	BODY HEALTH 1 • Microorganisms • Bacterial infections • Viral infections • Other infections
Unit 6	WEATHER • Elements of weather • Air masses and clouds • Fronts and storms • Weather forecasting	ENERGY 2 • Magnetism • Current and static electricity • Using electricity • Energy sources	BODY HEALTH 2 • Body defense mechanisms • Treating disease • Preventing disease • Community health
Unit 7	CLIMATE • Climate and weather • Worldwide climate • Regional climate • Local climate	MACHINES 1 • Measuring distance • Force • Laws of Newton • Work	ASTRONOMY • Extent of the universe • Constellations • Telescopes • Space explorations
Unit 8	HUMAN ANATOMY 1 • Cell structure and function • Skeletal and muscle systems • Skin • Nervous system	MACHINES 2 • Friction • Levers • Wheels and axles • Inclined planes	OCEANOGRAPHY • History of oceanography • Research techniques • Geology of the ocean • Properties of the ocean
Unit 9	HUMAN ANATOMY 2 • Respiratory system • Circulatory system • Digestive system • Endocrine system	BALANCE IN NATURE • Photosynthesis • Food • Natural cycles • Balance in nature	SCIENCE AND TOMORROW • The land • Waste and ecology • Industry and energy • New frontiers
Unit 10	CAREERS IN SCIENCE • Scientists at work • Astronomy • Meteorology • Medicine	SCIENCE AND TECHNOLOGY • Basic science • Physical science • Life science • Vocations in science	SCIENTIFIC APPLICATIONS • Measurement • Practical health • Geology and astronomy • Solving problems

SCIENCE SCOPE & SEQUENCE

Biology (Grade 10)	Chemistry (Grade 11)	Physics (Grade 12)	
TAXONOMY • History of taxonomy • Binomial nomenclature • Classification • Taxonomy	ESTIMATE AND MEASUREMENT • Metric units and instrumentation • Observation and hypothesizing • Scientific notation • Careers in chemistry	KINEMATICS • Scalars and vectors • Length measurement • Acceleration • Fields and models	Unit 1
BASIS OF LIFE • Elements and molecules • Properties of compounds • Chemical reactions • Organic compounds	ELEMENTS, COMPOUNDS, AND MIXTURES • Alchemy • Elements • Compounds • Mixtures	DYNAMICS • Newton's laws of motion • Gravity • Circular motion • Kepler's laws of planetary motion	Unit 2
MICROBIOLOGY • The microscope • Protozoan • Algae • Microorganisms	GASES AND MOLES • Kinetic theory • Gas laws • Combined gas law • Moles	WORK AND ENERGY • Mechanical energy • Conservation of energy • Power and efficiency • Heat energy	Unit 3
CELLS • Cell theories • Examination of the cell • Cell design • Cells in organisms	ATOMIC MODELS • Historical models • Modern atomic structure • Periodic Law • Nuclear reactions	WAVES • Energy transfers • Reflection and refraction of waves • Diffraction and interference • Sound waves	Unit 4
PLANTS: GREEN FACTORIES • The plant cell • Anatomy of the plant • Growth and function of plants • Plants and people	CHEMICAL FORMULAS • Ionic charges • Electronegativity • Chemical bonds • Molecular shape	LIGHT • Speed of light • Mirrors • Lenses • Models of light	Unit 5
HUMAN ANATOMY AND PHYSIOLOGY • Digestive and excretory system • Respiratory and circulatory system • Skeletal and muscular system • Body control systems	CHEMICAL REACTIONS • Detecting reactions • Energy changes • Reaction rates • Equilibriums	STATIC ELECTRICITY • Nature of charges • Transfer of charges • Electric fields • Electric potential	Unit 6
GENETICS AND INHERITANCE • Gregor Mendel's experiments • Chromosomes and heredity • Molecular genetics • Human genetics	EQUILIBRIUM SYSTEMS • Solutions • Solubility equilibriums • Acid-base equilibriums • Redox equilibriums	ELECTRIC CURRENTS • Electromotive force • Electron flow • Resistance • Circuits	Unit 7
CELL DIVISION AND REPRODUCTION • Mitosis and meiosis • Asexual reproduction • Sexual reproduction • Plant reproduction	HYDROCARBONS • Organic compounds • Carbon atoms • Carbon bonds • Saturated and unsaturated	MAGNETISM • Fields • Forces • Electromagnetism • Electron beams	Unit 8
ECOLOGY AND ENERGY • Ecosystems • Communities and habitats • Pollution • Energy	CARBON CHEMISTRY • Saturated and unsaturated • Reaction types • Oxygen groups • Nitrogen groups	ATOMIC AND NUCLEAR PHYSICS • Electromagnetic radiation • Quantum theory • Nuclear theory • Nuclear reaction	Unit 9
APPLICATIONS OF BIOLOGY • Principles of experimentation • Principles of reproduction • Principles of life • Principles of ecology	ATOMS TO HYDROCARBONS • Atoms and molecules • Chemical bonding • Chemical systems • Organic chemistry	KINEMATICS TO NUCLEAR PHYSICS • Mechanics • Wave motion • Electricity • Modern physics	Unit 10

STRUCTURE OF THE LIFEPAC CURRICULUM

The LIFEPAC curriculum is conveniently structured to provide one Teacher's Guide containing teacher support material with answer keys and ten student worktexts for each subject at grade levels 2 through 12. The worktext format of the LIFEPACs allows the student to read the textual information and complete workbook activities all in the same booklet. The easy-to-follow LIFEPAC numbering system lists the grade as the first number(s) and the last two digits as the number of the series. For example, the Language Arts LIFEPAC at the 6th grade level, 5th book in the series would be LAN0605.

Each LIFEPAC is divided into three to five sections and begins with an introduction or overview of the booklet as well as a series of specific learning objectives to give a purpose to the study of the LIFEPAC. The introduction and objectives are followed by a vocabulary section which may be found at the beginning of each section at the lower levels or in the glossary at the high school level. Vocabulary words are used to develop word recognition and should not be confused with the spelling words introduced later in the LIFEPAC. The student should learn all vocabulary words before working the LIFEPAC sections to improve comprehension, retention, and reading skills.

Each activity or written assignment in grades 2 through 12 has a number for easy identification, such as 1.1. The first number corresponds to the LIFEPAC section and the number to the right of the decimal is the number of the activity.

Teacher checkpoints, which are essential to maintain quality learning, are found at various locations throughout the LIFEPAC. The teacher should check 1) neatness of work and penmanship, 2) quality of understanding (tested with a short oral quiz), 3) thoroughness of answers (complete sentences and paragraphs, correct spelling, etc.), 4) completion of activities (no blank spaces), and 5) accuracy of answers as compared to the answer key (all answers correct).

The self test questions in grades 2 through 12 are also number coded for easy reference. For example, 2.015 means that this is the 15th question in the self test of Section 2. The first number corresponds to the LIFEPAC section, the zero indicates that it is a self test question, and the number to the right of the zero the question number.

The LIFEPAC test is packaged at the center of each LIFEPAC. It should be removed and put aside before giving the booklet to the student for study.

Answer and test keys in grades 2 through 12 have the same numbering system as the LIFEPACs. The student may be given access to the answer keys (not the test keys) under teacher supervision so that he can score his own work.

A thorough study of the Scope & Sequence by the teacher before instruction begins is essential to the success of the student. The teacher should become familiar with expected skill mastery and understand how these grade-level skills fit into the overall skill development of the curriculum. The teacher should also preview the objectives that appear at the beginning of each LIFEPAC for additional preparation and planning.

TEST SCORING AND GRADING

Answer keys and test keys give examples of correct answers. They convey the idea, but the student may use many ways to express a correct answer. The teacher should check for the essence of the answer, not for the exact wording. Many questions are high level and require thinking and creativity on the part of the student. Each answer should be scored based on whether or not the main idea written by the student matches the model example. "Any Order" or "Either Order" in a key indicates that no particular order is necessary to be correct.

Most self tests and LIFEPAC tests at the lower elementary levels are scored at 1 point per answer; however, the upper levels may have a point system awarding 2 to 5 points for various answers or questions. Further, the total test points will vary; they may not always equal 100 points. They may be 78, 85, 100, 105, etc.

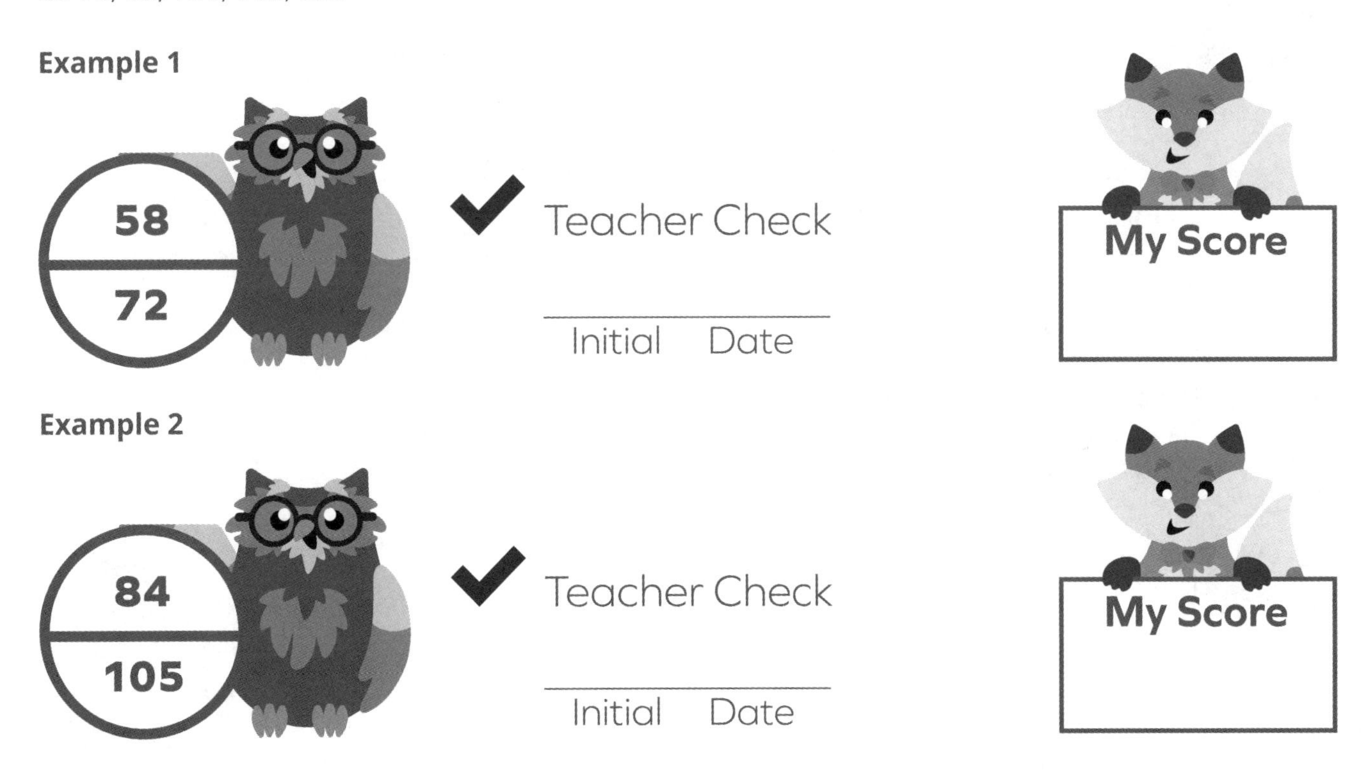

A score box similar to ex. 1 above is located at the end of each self test and on the front of the LIFEPAC test. The bottom score, 72, represents the total number of points possible on the test. The upper score, 58, represents the number of points your student will need to receive an 80% or passing grade. If you wish to establish the exact percentage that your student has achieved, find the total points of his correct answers and divide it by the bottom number (in this case 72). For example, if your student has a point total of 65, divide 65 by 72 for a grade of 90%. Referring to ex. 2, on a test with a total of 105 possible points, the student would have to receive a minimum of 84 correct points for an 80% or passing grade. If your student has received 93 points, simply divide the 93 by 105 for a percentage grade of 89%. Students who receive a score below 80% should review the LIFEPAC and retest using the appropriate Alternate Test found in the Teacher's Guide.

The following is a guideline to assign letter grades for completed LIFEPACs based on a maximum total score of 100 points.

Example:

LIFEPAC Test	=	60% of the Total Score (or percent grade)
Self Test	=	25% of the Total Score (average percent of self tests)
Reports	=	10% or 10* points per LIFEPAC
Oral Work	=	5% or 5* points per LIFEPAC

*Determined by the teacher's subjective evaluation of the student's daily work.

Example:

LIFEPAC Test Score	=	92%	92 × .60	=	55 points
Self Test Average	=	90%	90 × .25	=	23 points
Reports				=	8 points
Oral Work				=	4 points
TOTAL POINTS				=	90 points

Grade Scale based on point system:

100 – 94	=	A
93 – 86	=	B
85 – 77	=	C
76 – 70	=	D
Below 70	=	F

TEACHER HINTS AND STUDYING TECHNIQUES

LIFEPAC Activities are written to check the level of understanding of the preceding text. The student may look back to the text as necessary to complete these activities; however, a student should never attempt to do the activities without reading (studying) the text first. Self tests and LIFEPAC tests are never open book tests.

Language arts activities (skill integration) often appear within other subject curriculum. The purpose is to give the student an opportunity to test his skill mastery outside of the context in which it was presented.

Writing complete answers (paragraphs) to some questions is an integral part of the LIFEPAC Curriculum in all subjects. This builds communication and organization skills, increases understanding and retention of ideas, and helps enforce good penmanship. Complete sentences should be encouraged for this type of activity. Obviously, single words or phrases do not meet the intent of the activity, since multiple lines are given for the response.

Review is essential to student success. Time invested in review where review is suggested will be time saved in correcting errors later. Self tests, unlike the section activities, are closed book. This procedure helps to identify weaknesses before they become too great to overcome. Certain objectives from self tests are cumulative and test previous sections; therefore, good preparation for a self test must include all material studied up to that testing point.

The following procedure checklist has been found to be successful in developing good study habits in the LIFEPAC curriculum.

1. Read the introduction and Table of Contents.
2. Read the objectives.
3. Recite and study the entire vocabulary (glossary) list.
4. Study each section as follows:
   a. Read the introduction and study the section objectives.
   b. Read all the text for the entire section, but answer none of the activities.
   c. Return to the beginning of the section and memorize each vocabulary word and definition.
   d. Reread the section, complete the activities, check the answers with the answer key, correct all errors, and have the teacher check.
   e. Read the self test but do not answer the questions.
   f. Go to the beginning of the first section and reread the text and answers to the activities up to the self test you have not yet done.
   g. Answer the questions to the self test without looking back.
   h. Have the self test checked by the teacher.
   i. Correct the self test and have the teacher check the corrections.
   j. Repeat steps a–i for each section.
5. Use the SQ3R method to prepare for the LIFEPAC test.

   Scan the whole LIFEPAC.
   Question yourself on the objectives.
   Read the whole LIFEPAC again.
   Recite through an oral examination.
   Review weak areas.
6. Take the LIFEPAC test as a closed book test.
7. LIFEPAC tests are administered and scored under direct teacher supervision. Students who receive scores below 80% should review the LIFEPAC using the SQ3R study method and take the Alternate Test located in the Teacher's Guide. The final test grade may be the grade on the Alternate Test or an average of the grades from the original LIFEPAC test and the Alternate Test.

GOAL SETTING AND SCHEDULES

Each school must develop its own schedule, because no single set of procedures will fit every situation. The following is an example of a daily schedule that includes the five LIFEPAC subjects as well as time slotted for special activities.

Possible Daily Schedule

8:15	-	8:25	Pledges, prayer, songs, devotions, etc.
8:25	-	9:10	Bible
9:10	-	9:55	Language Arts
9:55	-	10:15	Recess (juice break)
10:15	-	11:00	Math
11:00	-	11:45	History & Geography
11:45	-	12:30	Lunch, recess, quiet time
12:30	-	1:15	Science
1:15	-		Drill, remedial work, enrichment*

***Enrichment:** *Computer time, physical education, field trips, fun reading, games and puzzles, family business, hobbies, resource persons, guests, crafts, creative work, electives, music appreciation, projects.*

Basically, two factors need to be considered when assigning work to a student in the LIFEPAC curriculum.

The first is time. An average of 45 minutes should be devoted to each subject, each day. Remember, this is only an average. Because of extenuating circumstances, a student may spend only 15 minutes on a subject one day and the next day spend 90 minutes on the same subject.

The second factor is the number of pages to be worked in each subject. A single LIFEPAC is designed to take three to four weeks to complete. Allowing about three to four days for LIFEPAC introduction, review, and tests, the student has approximately 15 days to complete the LIFEPAC pages. Simply take the number of pages in the LIFEPAC, divide it by 15 and you will have the number of pages that must be completed on a daily basis to keep the student on schedule. For example, a LIFEPAC containing 45 pages will require three completed pages per day. Again, this is only an average. While working a 45-page LIFEPAC, the student may complete only one page the first day if the text has a lot of activities or reports, but go on to complete five pages the next day.

Long-range planning requires some organization. Because the traditional school year originates in the early fall of one year and continues to late spring of the following year, a calendar should be devised that covers this period of time. Approximate beginning and completion dates can be noted on the calendar as well as special occasions such as holidays, vacations and birthdays. Since each LIFEPAC takes three to four weeks or 18 days to complete, it should take about 180 school days to finish a set of ten LIFEPACs. Starting at the beginning school date, mark off 18 school days on the calendar and that will become the targeted completion date for the first LIFEPAC. Continue marking the calendar until you have established dates for the remaining nine LIFEPACs making adjustments for previously noted holidays and vacations. If all five subjects are being used, the ten established target dates should be the same for the LIFEPACs in each subject.

TEACHING SUPPLEMENTS

The sample weekly lesson plan and student grading sheet forms are included in this section as teacher support materials and may be duplicated at the convenience of the teacher.

The student grading sheet is provided for those who desire to follow the suggested guidelines for assignment of letter grades as previously discussed. The student's self test scores should be posted as percentage grades. When the LIFEPAC is completed, the teacher should average the self test grades, multiply the average by .25, and post the points in the box marked self test points. The LIFEPAC percentage grade should be multiplied by .60 and posted. Next, the teacher should award and post points for written reports and oral work. A report may be any type of written work assigned to the student whether it is a LIFEPAC or additional learning activity. Oral work includes the student's ability to respond orally to questions which may or may not be related to LIFEPAC activities or any type of oral report assigned by the teacher. The points may then be totaled and a final grade entered along with the date that the LIFEPAC was completed.

The Student Record Book which was specifically designed for use with the Alpha Omega curriculum provides space to record weekly progress for one student over a nine-week period as well as a place to post self test and LIFEPAC scores. The Student Record Books are available through the current Alpha Omega catalog; however, unlike the enclosed forms these books are not for duplication and should be purchased in sets of four to cover a full academic year.

WEEKLY LESSON PLANNER

Week of:

Monday	Subject	Subject	Subject	Subject
Tuesday	Subject	Subject	Subject	Subject
Wednesday	Subject	Subject	Subject	Subject
Thursday	Subject	Subject	Subject	Subject
Friday	Subject	Subject	Subject	Subject

WEEKLY LESSON PLANNER

Week of:

	Subject	Subject	Subject	Subject
Monday				
	Subject	Subject	Subject	Subject
Tuesday				
	Subject	Subject	Subject	Subject
Wednesday				
	Subject	Subject	Subject	Subject
Thursday				
	Subject	Subject	Subject	Subject
Friday				

Student Name ______________________ Year ____________

Bible

LP	Self Test Scores by Sections 1	2	3	4	5	Self Test Points	LIFEPAC Test	Oral Points	Report Points	Final Grade	Date
01											
02											
03											
04											
05											
06											
07											
08											
09											
10											

History & Geography

LP	Self Test Scores by Sections 1	2	3	4	5	Self Test Points	LIFEPAC Test	Oral Points	Report Points	Final Grade	Date
01											
02											
03											
04											
05											
06											
07											
08											
09											
10											

Language Arts

LP	Self Test Scores by Sections 1	2	3	4	5	Self Test Points	LIFEPAC Test	Oral Points	Report Points	Final Grade	Date
01											
02											
03											
04											
05											
06											
07											
08											
09											
10											

Student Name ______________________ Year ____________

Math

LP	Self Test Scores by Sections 1	2	3	4	5	Self Test Points	LIFEPAC Test	Oral Points	Report Points	Final Grade	Date
01											
02											
03											
04											
05											
06											
07											
08											
09											
10											

Science

LP	Self Test Scores by Sections 1	2	3	4	5	Self Test Points	LIFEPAC Test	Oral Points	Report Points	Final Grade	Date
01											
02											
03											
04											
05											
06											
07											
08											
09											
10											

Spelling/Electives

LP	Self Test Scores by Sections 1	2	3	4	5	Self Test Points	LIFEPAC Test	Oral Points	Report Points	Final Grade	Date
01											
02											
03											
04											
05											
06											
07											
08											
09											
10											

INSTRUCTIONS FOR SCIENCE

The LIFEPAC curriculum from grades 2 through 12 is structured so that the daily instructional material is written directly into the LIFEPACs. The student is encouraged to read and follow this instructional material in order to develop independent study habits. The teacher should introduce the LIFEPAC to the student, set a required completion schedule, complete teacher checks, be available for questions regarding both content and procedures, administer and grade tests, and develop additional learning activities as desired. Teachers working with several students may schedule their time so that students are assigned to a quiet work activity when it is necessary to spend instructional time with one particular student.

The Teacher Notes section of the Teacher's Guide lists the required or suggested materials for the LIFEPACs and provides additional learning activities for the students. The materials section refers only to LIFEPAC materials and does not include materials which may be needed for the additional activities. Additional learning activities provide a change from the daily school routine, encourage the student's interest in learning, and may be used as a reward for good study habits.

SCIENCE 201

Unit 1: The Living and Nonliving

TEACHER NOTES

MATERIALS NEEDED FOR LIFEPAC	
Required	
• bean seeds • paper cups • soil	• 1 rock, 2 cups • 2 labels, 3 seeds

ADDITIONAL LEARNING ACTIVITIES

Section 1: God Made Living and Nonliving Objects

1. Discuss these questions with the students:
   a. What was the earth like after the first day?
   b. What would you have seen on Earth after the second day?
   c. Can you tell the differences between living and nonliving things?
   d. How does the sun help you to know what season it is?
   e. How are fish special for living in the water?
   f. How was man different from the other animals?
2. Invite a Christian to speak to the students about the Creation of the world.
3. Have the students draw a mural on a large sheet of butcher paper depicting the sequence of Creation.
4. Have students make a collection of living and nonliving objects.
5. Make up games using the homes of living and nonliving insects on cards. Each player that can correctly say the object is living or nonliving gets to move forward one space. If no special board is available, use any game board with a path on it. The cards can be made by these students with the whole class involved in playing the game.

Section 2: Rock and Seed Experiment

This section is a special study and does not require further activities.

Section 3: God Made Objects and Man Made Objects

1. Discuss these questions with the students:
   a. Can you see anything that God made?
   b. Can you see anything that man made?
   c. What is the difference in the things God made and in the things that man made?
   d. What does man use to make objects?

2. To help students to become aware of different materials, students can describe objects stressing the qualities of softness, hardness, shininess, dullness, pliability, and so forth.
3. Make a list of all objects the students see. List the materials used in each object. Teach categories as you progress with this discussion.
4. Students may construct objects from different materials.

Administer the LIFEPAC Test.

The test is to be administered in one session. Give no help except with directions.
Evaluate the tests and review areas where the students have done poorly.
Review the pages and activities that stress the concepts tested.
If necessary, administer the Alternate LIFEPAC test.

ANSWER KEYS

SECTION 1

1.1 God
1.2 daytime
1.3 nighttime
1.4 Any order:
a. day b. night
1.5 sky
1.6 Any order:
a. dry land or earth
b. sea or water
1.7 Any order:
a. plants or grass
b. trees
1.8 Answers will vary.
1.9 Teacher check
1.10 a. water b. air
c. light d. food
1.11 they are not living
1.12 Answers will vary.
1.13 Teacher check
1.14 Any order:
a. sun b. moon
1.15 to light the earth that night
1.16 the days
1.17 Any order:
a. summer b. fall
c. winter d. spring
1.18 Any order:
a. sun b. moon
c. stars
1.19 Any order:
a. fish b. birds
1.20 Any order:
a. air b. light
c. water d. food
1.21 Any order:
a. animals b. cattle
c. reptiles d. man
1.22 man
1.23 He made fish and birds.
1.24 He made animals, cattle, reptiles, and man.
1.25 He rested.
1.26 Teacher check
1.27 c
1.28 b
1.29 g
1.30 f
1.31 e
1.32 d

SELF TEST 1

1.01-1.06 Any order:
1.01 a. day b. night
1.02 a. sky
1.03 a. earth b. sea
1.04 a. sun b. moon
c. stars
1.05 a. fish b. birds
1.06 a. animals b. reptiles
c. cattle d. man
1.07 to make something
1.08 the world we live on
1.09 set apart from other parts
1.010 heaven

1.011 Any order:
a. air b. water
c. food d. light
1.012 not alive
1.013-1.014 Any order:
1.013 a. sun b. moon
1.014 a. light b. food
c. water d. air
1.015 God

SECTION 2

2.1 I planted my rock.
2.2 My rock is in the ground.
2.3 I watered my rock.
2.4 My rock is in the ground.
2.5 I watered my rock.
I put my rock in the sun.
2.6 My rock is in the ground.
2.7 I watered my rock.
I put plant food on my rock.
2.8 My rock is in the ground.
2.9 I watered my rock.
2.10 My rock is in the ground.
2.11 I planted my seeds.
2.12 My seeds are in the ground.
2.13 I watered my seeds.
2.14 My seeds are in the ground.
2.15 I watered my seeds.
I put my seeds in the sun.
2.16 My seeds are in the ground.
2.17 I watered my seeds.
I put plant food on my seeds.
2.18 My seeds are in the ground.
2.19 I watered my seeds.
2.20 A plant is growing.
2.21 sun
2.22 run
2.23 such
2.24 trust
2.25 dust
2.26 Teacher check

SELF TEST 2

2.01 plant (seed)
2.02 rock
2.03 Any order:
a. light b. water
c. air d. food
2.04 God
2.05 God
2.06 yes
2.07 no
2.08 yes
2.09 yes
2.010 no

SECTION 3

3.1-3.8 Answers will vary.
3.1 hill, beach, river, mountain
3.2
1. desk
2. picture
3. clock
4. pencil
5. paper

3.3
1. scissors
2. eraser
3. chalk
4. gloves
5. boots

3.4 a. lawn mower b. pan c. spoon d. fork
3.5 wood, plastic
3.6 plastic, wood, metal
3.7 metal, glass
3.8 wood, metal, plastic
3.9 desk, tent, chest, shelf

SELF TEST 3

3.01 God
3.02 man
3.03 grow
3.04 wood, plastic
3.05 glass
3.06 metal, glass, plastic, wood
3.07 paper, plastic
3.08 plastic, metal, wood
3.09-3.016 Answers will vary.
3.017-3.020 Any order:
3.017 light
3.018 food
3.019 air
3.020 water
3.021 c
3.022 d
3.023 a
3.024 e
3.025 b

LIFEPAC TEST

1. nonliving
2. nonliving
3. nonliving
4. living
5. nonliving
6. living
7. living
8. nonliving
9. nonliving
10. nonliving
11. air
12. light
13. water
14. food
15. It does not grow.
16. man
17. moon
18. sea
19. star
20. sun
21. pen
22. car
23. desk
24. book
25. toy
26. metal
27. glass
28. wood, metal, or plastic
29. wood
30. plastic, metal, or wood

ALTERNATE LIFEPAC TEST

1.-5. Examples; any order:
1. desk
2. pencil
3. crayon
4. paper
5. book
6.-10. Examples; any order:
6. dog
7. cat
8. tomato plant
9. tree
10. boy
11. circle day
12. X on desk
13. circle sky
14. circle man
15. X on paper
16. circle tree
17. X on book
18. circle star
19. X on pen
20. X on chair
21. X on rug
22. circle night
23. circle sea
24. circle water
25.-28. Any order:
25. light
26. water
27. food
28. air

SCIENCE 201

ALTERNATE LIFEPAC TEST

Name ______________________

Date ______________________

Each answer = 1 point

Write the names of five things that are nonliving.

1. ______________________

2. ______________________

3. ______________________

4. ______________________

5. ______________________

Write the names of five things that are living.

6. ______________________

7. ______________________

8. ______________________

9. ______________________

10. ______________________

Circle the things that God created. Put an X on things man has made.

11. day

12. desk

13. sky

14. man

15. paper

16. tree

17. book

18. star

19. pen

20. chair

21. rug

22. night

23. sea

24. water

Write the four things that living objects need to live.

25. ______________________________

26. ______________________________

27. ______________________________

28. ______________________________

SCIENCE 202

Unit 2: Plants

TEACHER NOTES

MATERIALS NEEDED FOR LIFEPAC	
Required	
• sweet potato	• rich soil
• 3 toothpicks	• egg carton
• 2 jars of water	• teaspoon
• celery stalks	• 10 half-eggshells
• food coloring	• 10 bean seeds, water

ADDITIONAL LEARNING ACTIVITIES

Section 1: How Plants Are Like People

1. Discuss these questions with the students:
   a. How does God give the earth water?
   b. Does the air smell different when it rains?
   c. Why do you like having trees and flowers in your yard?
2. Have students, during a rainstorm, set a bucket outside. Measure the amount of water in the bucket.
3. Have the students collect pictures of ways people pollute the air: traffic, garbage, sprays, smoke, and so forth.
4. Have a student draw pictures showing how plants help people.
5. Have the student draw pictures showing how plants and people are alike.

Section 2: How Plants Are the Same

1. Discuss these questions with the students:
   a. What do you think makes a tiny seed grow into a big plant?
   b. What is in the veins of a person? What is in the veins of a leaf?
   c. Do people use plants for anything other than food?
2. Have the students make a booklet of the shapes and sizes of different seeds.
3. As a demonstration, put an avocado seed that has been peeled and slightly split into a jar of water. Put toothpicks into the sides of the seed to keep the larger half out of the water. Watch the root begin to grow.
4. Fill a pot with soil. Have the students bring a grapefruit seed from breakfast each day. Every day, plant one seed starting in the center and making a spiral out. If each student participates they will have a tree of their own.

Section 3: How Plants Are Different

1. Discuss these questions with the students:
    a. Why do you think God made different kinds of plants?
    b. Can you tell about a strange plant you have seen?
    c. How do plants get their names? Can you think of other names you might give certain plants?
2. Have a group of students collect pictures of different kinds of habitats such as swamps, deserts, and so forth.
3. Have the students take care of plants at school by watering and feeding them on a regular schedule.
4. As a class activity, collect pictures and categorize trees as evergreens or as trees that lose their leaves.

Section 4: What Plants Need to Live

1. Discuss these questions with the students:
    a. How do your parents bring fresh air into the house?
    b. What do you think would happen if we did not have the sun?
2. Assign the students to make a chart showing how many hours of sunlight are in a summer day and in a winter day. Draw conclusions about when the plants grow the most.
3. Have the students fill three pots with the same kind of soil and plant the same kind of seeds in them. Place them on the same windowsill. Keep one plant moist and the soil loose to let in air. Keep the second plant dry with the soil hard-packed. Keep the third flooded with water. See which plant grows best.
4. Assign the students to punch a hole through a piece of cardboard and slip the stem of a leaf through the hole. Set the stem in a glass of water. Put an empty glass over the top of the leaf resting on the cardboard. Moisture will gather on the leaf after a few hours.

Administer the LIFEPAC Test.

The test is to be administered in one session. Give no help except with directions.
Evaluate the tests and review areas where the students have done poorly.
Review the pages and activities that stress the concepts tested.
If necessary, administer the Alternate LIFEPAC test.

ANSWER KEYS

SECTION 1

1.1 The sun makes the land warm.
1.2 Water is to drink.
1.3 Air is needed by plants and people.
1.4 Ground is for work and play.

SELF TEST 1

1.01 sun
1.02 people
1.03 work
1.04 live
1.05 water
1.06 grow
1.07 shine
1.08 alike

SECTION 2

2.1 d. are alike in some ways
2.2 e. have tiny plants in them
2.3 c. grow up from the seed
2.4 a. grow down into the ground
2.5 b. grow on the stem
2.6 window, into, begin, fin
2.7 leaves
2.8 veins
2.9 Air
2.10 green
2.11 Tubes
2.12 job
2.13 fall
2.14 roots
2.15 spring
2.16 leaves
2.17 branches
2.18 summer
2.19-2.20 Answers will vary.
2.19 Curled-up leaves are in the leaf bud.
2.20 Curled-up flowers are in the flower bud.
2.21 blossoms
2.22 spring
2.23 seeds
2.24 soft
2.25 flower
2.26 hard

SELF TEST 2

2.01 3
2.02 1
2.03 2
2.04 4
2.05 root
2.06 stem
2.07 leaves
2.08 water
2.09 bottom
2.010 root
2.011 seeds
2.012 soft
2.013 Nuts
2.014 air
2.015 veins

SECTION 3

3.1 habitat
3.2 Cactus
3.3 winter
3.4 warm
3.5 warm
3.6 cold
3.7 A tulip grows from a bulb.
3.8 A root grows down into the ground.
3.9 A cutting can grow roots.
3.10 Some plants must be planted again every year.
3.11 The water lily comes on top of the water in the morning.
3.12 A sunflower always faces the sun.
3.13 A fly trap shuts around a fly.

SELF TEST 3

3.01 habitats
3.02 water
3.03 Orchids
3.04 roots
3.05 Air
3.06 does make food.
3.07 does not make food
3.08 is a stem and leaves
3.09 carry water
3.010 grow from cuttings
3.011-3.017 Answers will vary.
3.011 Sunshine helps plants to grow.
3.012 A stem grows up from the ground.
3.013 A plant will grow from a bulb.
3.014 A seed can grow into a plant.
3.015 A sunflower always faces the sun.
3.016 habitat, The place where a plant grows well is its habitat.
3.017 growth, Plants grow in different ways.

SECTION 4

4.1 ponds gone
4.2 fox boxer
4.3 often offer
4.4 green
4.5 sunshine
4.6 rest
4.7 roots
4.8 sunshine
4.9 light
4.10 rest
4.11 Plants take in water by their roots.
4.12 Plants need air.
4.13 Sunshine helps plants to grow.

SELF TEST 4

4.01 The green of the leaf makes food.
4.02 Plants need a rest at night.
4.03 Plants take in water through their roots.
4.04 hard
4.05 nuts
4.06 brown
4.07 seeds
4.08 water
4.09 leaves
4.010 roots
4.011 stem
4.012 yes
4.013 no
4.014 no
4.015 no

LIFEPAC TEST

1. a. flower or blossom
b. buds
c. leaf
d. fruit
e. seed
f. root
g. stem

2. seed, roots, stem, leaves

3. a. air
b. water
c. sunshine

4. a. habitat
b. growth

5. green

6. tubes

7. rest

8. sunflower

9. veins

10. cactus

11. ground

12. fly trap

13. geranium

14. tulip

15. I grew a sweet potato.
It grew leaves and roots
It is pretty.

ALTERNATE LIFEPAC TEST

1. yes

2. yes

3. yes

4. yes

5. yes

6. no

7. no

8. yes

9. yes

10. yes

11. job

12. leaves

13. God

14. lily

15. blossom

16. sunflower

17. water

18. fly trap

19. seeds

20. lily

SCIENCE 202

ALTERNATE LIFEPAC TEST

Name ______________________

Date ______________________

Circle *Yes* or *No*.

1. God made people and plants alike in many ways.

Yes No

2. Plants need air, sunshine, and water to live.

Yes No

3. The roots hold the plant in the ground.

Yes No

4. You plant a seed when it is ripe.

Yes No

5. Some plants have seeds, but some have bulbs.

Yes No

6. No plant can grow from a cutting.

Yes No

7. All plants grow year after year.

Yes No

8. The orchid is very different from the cactus.

Yes No

9. The habitat is where a plant lives.

Yes No

10. Water and food move in the stem.

Yes No

Write the word from the box on the correct line.

leaves	God	lily	job	blossom

11. We all have one to do. ______________________

12. They grow on the stem. ______________________

13. He made everything. ______________________

14. It lives in the water. ______________________

15. It makes a flower pretty. ______________________

Choose the best word and write it in the sentence.

water	seeds	lily	fly trap	sunflower

16. The flower that follows the sun is the ______________ .

17. The stem carries food and ______________ to the leaves.

18. The plant that eats flies is called the ______________ .

19. The main job of the flower is to make ______________ .

20. A flower that goes underwater at night is the __________ .

SCIENCE 203

Unit 3: Animals

TEACHER NOTES

MATERIALS NEEDED FOR LIFEPAC	
Required	Suggested
• glass jar • dry grass or hay • water • microscope	• Bible, King James version

ADDITIONAL LEARNING ACTIVITIES

Section 1: How Animals Are Alike

1. Show pictures of animals and ask:
   - Can you name these animals? (Write animal names on the board.)
   - Can you think of any ways all of these animals are alike?
   - One way they are alike is that they all have keen senses. Can you name your five senses? Explain that all animals have senses. Some have keener senses than people. For example, an eagle can see a mouse a mile away. A dog can hear sounds that people cannot hear, and so forth.
2. Write the word *communicate* on the chalkboard. Tell students, "People communicate by talking to each other. Can you think of ways that animals communicate?" (dogs growl, birds chirp, etc.)
3. Write the word *clean* on the chalkboard. Tell students, "Another way animals are alike is that they try to be clean. People keep clean by washing and taking baths. Can you think of ways that animals keep clean?" (Discuss bird baths, cats licking their fur, etc.)
4. Divide students into groups of three. Assign each group one animal. Each child should draw or cut out a picture of that animal. One student should tell about the keen senses the animal has. Another student could tell how the animal communicates, and a third student could tell how the animal keeps clean.
5. Assign the students to read a book about animals and share one interesting fact with the class.
6. Individually or as a class, fold a piece of paper in half lengthwise. Use the fold line to divide the paper into two columns. In one column, name an animal. In the second column, tell the way the animal communicates, that is, *bird chirps* and so forth.

Section 2: How Animals Are Different

1. Discuss these questions with the students:
   a. We have learned the ways animals are alike. Can you think of ways that animals are different?

   b. Do all animals look alike?
   c. Are animals the same sizes and shapes? (Discuss appearance of animals with and without backbones: worm and squid versus dog, cat, and so forth.)
   d. Do all animals live in the same kinds of homes?

2. Have each student bring one animal picture to school. Set up two empty boxes marked "Backbone" and "No Backbone." As each child holds up a picture and names the animal, have them place it in the appropriate box. Repeat this activity using the words "Large" and "Small" on the boxes.
3. Have each student bring a picture of an animal to school. Have each child show his picture, name the animal, and tell where it makes its home.
4. Assign the students to draw a picture of an animal in its home.
5. Individually or as a class, fold a large piece of drawing paper in half. On one side, write "Backbone." On the other side, write "No Backbone." Draw appropriate pictures on each side of the paper.

Section 3: What Animals Need to Live

1. Discuss these questions with the students:
   a. What do people need to live? (air, water, food)
   b. What things do animals need to live? (same as people)
   c. People and animals also need to be safe. Can you think of a way a porcupine keeps safe? a dog? a bee?
   d. Instinct also keeps animals safe. Instinct is knowing how to do something without learning how, like when birds fly south for the winter. Can you think of other ways animals use instinct to stay safe?
2. Discuss how animals find protection from storms. Introduce the story of Noah and his ark.
3. Have groups or the whole class make a chart together. Use the following headings.

Animal Picture	Animal Name	Animal Eats

4. Assign the students to draw a picture of an animal showing how it protects itself.
5. Have the students construct an ark out of paper or cardboard. Put toy animals, or pictures of animals in the ark. Use the ark to tell the story of Noah.

Administer the LIFEPAC Test.

The test is to be administered in one session. Give no help except with directions.
Evaluate the tests and review areas where the students have done poorly.
Review the pages and activities that stress the concepts tested.
If necessary, administer the Alternate LIFEPAC test.

ANSWER KEYS

SECTION 1

1.1 Any order:

Smell	**Taste**
cupcake	cupcake
roses	lemonade
smoke	
wet socks	
rain	

See	**Hear**	**Touch**
bells	bells	cactus
cactus	horns	roses
cupcake	rain	trees
smoke		velvet
sun		wet socks
trees		bells
roses		horns
velvet		
wet socks		

1.2 hear, feel, smell, see, taste
1.3 senses
1.4 keen
1.5 survive
1.6 eagles
1.7 dogs
1.8 porpoises
1.9 fish
1.10 moles
1.11 squeak
1.12 need
1.13 eagle
1.14 clean
1.15 beaver
1.16 bees
1.17 smell
1.18 taste
1.19 touch
1.20 hear
1.21 see

SELF TEST 1

1.01 senses
1.02 senses
1.03 communicate
1.04 clean
1.05 five
1.06 touch
1.07 hearing
1.08 sight
1.09 taste
1.010 smell
1.011 smell, see, hear, feel, taste

SECTION 2

2.1 Teacher check
2.2 backbones
2.3 earthworms
2.4 senses
2.5 taste, best
2.6 beaver
2.7 gorilla
2.8 bear
2.9 bird
2.10 prairie dog
2.11 woodpecker
2.12 a. holes
b. polar bear
c. camel
d. nest
e. snake
2.13 shell
2.14 feathers
2.15 tough skin
2.16 scales
2.17 fur

SELF TEST 2

2.01 backbone
2.02 caves
2.03 white
2.04 desert
2.05 scales
2.06 feathers
2.07 yes
2.08 yes
2.09 no
2.010 yes
2.011 yes
2.012 no

SECTION 3

3.1 lungs
3.2 gills
3.3 skin
3.4 meat or plants
3.5 plants or meat
3.6 meat, plants
3.7 water
3.8 safe
3.9 quills
3.10

rain	train	paint
tail	lay	away
gray	afraid	stay

3.11 middle
3.12 end
3.13 water
3.14 warm
3.15 color
3.16 Instinct
3.17 odor

SELF TEST 3

3.01 oxygen to live
3.02 their gills
3.03 food
3.04 see
3.05 no backbone
3.06 instinct
3.07 safe
3.08 communicate
3.09 nest
3.010 feel
3.011 smell
3.012 hear
3.013 taste
3.014 see
3.015 no
3.016 no
3.017 yes
3.018 no
3.019 yes
3.020 yes
3.021 no

LIFEPAC TEST

1. stay alive
2. earthworm
3. sharp
4. dog
5. send messages
6. taste
7. touch
8. see
9. hear
10. smell
11. senses
12. communicate
13. clean
14. sizes
15. homes
16. ark
17. food
18. instinct
19. see
20. people

ALTERNATE LIFEPAC TEST

1. sharp
2. send messages
3. stay alive
4. dog
5. earthworm
6. smell
7. hear
8. see
9. touch
10. taste
11. see
12. ark
13. clean
14. senses
15. communicate
16. sizes
17. people
18. instinct
19. food
20. homes

SCIENCE 203

ALTERNATE LIFEPAC TEST

Name ____________________

Date ____________________

My Score ______ 16/20

Each answer = 1 point

Draw lines to match the word to the meaning.

1.	keen ▸	◂ dog
2.	communicate ▸	◂ sharp
3.	survive ▸	◂ earthworm
4.	backbone ▸	◂ stay alive
5.	no backbone ▸	◂ send messages

Write the sense used on the correct line.

smell	taste	see	hear	touch

6. __________ Matt knew by the odor that a skunk was near.

7. __________ Bill listened to the song.

8. __________ Ann looked at her book.

9. __________ Bob felt the cold water.

10. __________ Tom likes cookies.

Print the word from the box on the correct line.

see	homes	people	ark	sizes
food	senses	clean	instinct	communicate

11. Some animals stay safe because they are hard to ________________ .

12. Noah put the animals in the ________________ .

13. Animals try to stay ________________ .

14. Animals are alike. They have keen ________________ .

15. Animals ________________ with each other.

16. Animals are different ________________ and shapes.

17. Both animals and ________________ can communicate.

18. Animals know how to survive. Their ________________ helps them.

19. Animals need air, ________________ , and water.

20. Animals live in different kinds of ________________ .

SCIENCE 204

Unit 4: You

TEACHER NOTES

MATERIALS NEEDED FOR LIFEPAC	
Required	
• crayons • plain drawing paper • lined writing paper	• construction paper • family photograph

ADDITIONAL LEARNING ACTIVITIES

Section 1: You Are Like Other People

1. Choose two children from the class to stand up and say:
   "Here are two people. They have different names and they look different. Can you think of anything that is alike about them?" Then ask, "What is alike about all people?" (Discuss bones, muscles, skin, and body organs.)
2. Have groups of students work together to make a large drawing of a person. Have them label all the body parts they have studied.
3. Have students choose one body organ. Instruct them to draw a picture of the organ and give an oral or written report of the organ's function.

Section 2: You Are Different from Other People

1. Show pictures of two different people to the class. Discuss the ways the two people look different.
2. Discuss ways two people may act or think differently. Identical twins would make a good example.
3. Discuss how two people may think differently and like different things.
4. Have children cut out two pictures of people from magazines. Let them take turns holding up their pictures and telling how the people look different.
5. Play "I'm Thinking of Someone." The child who is it must describe another child in the group. For example: "I'm thinking of someone with brown hair and blue eyes. This person is tall and thin and loves to play piano." The other children in the group then guess the name of the person. The first child to guess correctly is then it.
6. Have students make a growing album. Students can start with a picture of them as a baby, one year old, and so forth. Have them show it to the class and explain how they have changed.
7. Have students draw a picture of themselves. Then, tell them to write a few sentences telling how they are different from other people and a few more telling about the things they like.

Section 3: You Are Part of a Family

1. Introduce the word *family*. "A family is a group of people you live with."
2. Have children list the members of their families. Say to students, "Families do things for each other. Can you name a person in your family and tell what that person does for you?" "Can you name a person in your family and tell what you do for that person?"
3. Discuss families. Say, "Families have fun together. Can you think of something you do with your family that is lots of fun?"
4. Have children group into pretend families. One child can be the mother, father, baby, brother, and so forth. Have them dramatize helping each other and having fun together.
5. Have students make a book about their family with the following directions: On each page, draw a picture of one family member. Under each picture, write a few sentences about how they and that person help each other and how they have fun together.

Section 4: You Need Good Health Habits

1. Discuss why it is important to be healthy. Let students describe being sick.
2. With the students, discuss ways of taking care of their body and keeping it healthy. Include good food, rest, bathing, brushing teeth, and so forth.
3. Groups of children can make a list of three or four good health habits. From that list they can make a chart with their names and the days of one week. Have them check off bathing, brushing teeth, and so on, each day until it becomes a habit.
4. Have the students cut out pictures of foods from magazines. Make a large poster. Draw a large circle on it with dividing lines and a little circle to the side to represent MyPlate. Have children paste their pictures in the correct group.
5. Have students keep a record of the food they eat for one day. Make five squares on a piece of paper. Write fruits in one square, vegetables in a second square, dairy in another square, grains in one, and protein in the last square. Write in each square the foods from each group they had that day. Discuss why certain foods are healthier than others.
6. Assign students to help their mother or father plan and make dinner. Be sure to include foods from all the sections of MyPlate.

Administer the LIFEPAC Test.

The test is to be administered in one session. Give no help except with directions.
Evaluate the tests and review areas where the students have done poorly.
Review the pages and activities that stress the concepts tested.
If necessary, administer the Alternate LIFEPAC test.

ANSWER KEYS

SECTION 1

1.1 Teacher check
1.2 Teacher check
1.3 God
1.4 bones
1.5 Muscles
1.6 Organs
1.7 skin
1.8 bones, muscles, skin, organs

SELF TEST 1

1.01 hard part inside your body
1.02 soft part inside your body that helps you move
1.03 part inside your body that does a certain job
1.04 take care of
1.05 organ used for thinking
1.06 organ that sends blood through your body
1.07 bones
1.08 muscles
1.09 skin
1.010 organs
1.011 God
1.012 move bones
1.013 skin
1.014 no
1.015 yes
1.016 job
1.017 bones
1.018 stomach
1.019 pores
1.020 God

SECTION 2

2.1 Teacher check
2.2 no
2.3 head, arms, feet, neck, fingers
2.4 Answers will vary.
Examples: color of hair, color of eyes, color of skin, height, weight
2.5

purple	blue
steak	steak
basketball	baseball
spelling	art
doctor	minister

2.6

a. 4	b. 2
c. 6	d. 5
e. 1	f. 3

2.7 color, shape
2.8 God wants people to be different.
2.9 yes
2.10 hook, hood, foot
2.11 broom, stool, tooth
2.12 **Book:** hook, foot, hood, cook, wool, stood
Balloon: tooth, broom, stool, moon, noon, goose, pool, tool

SELF TEST 2

2.01 like machines inside your body
2.02 part of your body where food goes after it is chewed
2.03 covers your body
2.04 the same
2.05 like the best
2.06 God
2.07 a. muscles
b. organs
c. bones
2.08 skin
2.09 thoughts
2.010 looks
2.011 growing
2.012 foot
2.013 shook
2.014 cookie
2.015 good
2.016 boot
2.017 room
2.018 school
2.019 tooth
2.020 room

SECTION 3

3.1-3.8 Answers will vary. Examples:
3.1 5
3.2 2
3.3 3
3.4 Dad
3.5 baby
3.6 camping, fishing, going to games, swimming, going to church
3.7 He makes sure we have food, clothes, and a place to live. He takes me to church. He fixes my bike.
3.8 She cooks, cleans, washes clothes, takes me to the library, takes me to the doctor.
3.9 Teacher check
3.10 teach
3.11 head
3.12 ea, ea, ea
3.13 ea, ea, ea
3.14 ea, ea, ea
3.15 ea, ea
3.16 ea, ea, ea
3.17 ea, ea

SELF TEST 3

3.01 a person who is fully grown
3.02 by doing things for each other
3.03 a set of the same kinds of things
3.04 a group of people who live together
3.05 like other people and different, too
3.06 feather
3.07 bread
3.08 ready
3.09 heavy
3.010 beach
3.011 reach
3.012 please
3.013 peach
3.014 foot
3.015 cook
3.016 stood
3.017 wool
3.018 tool
3.019 broom
3.020 moon
3.021 goose

SECTION 4

4.1 Teacher check
4.2 a. air b. food
c. water
4.3 vegetables, fruits, grains (whole grains), dairy (milk products—yogurt, cheese, cottage cheese, etc.), proteins
4.4 solid fats, salt, sweets or sugar
4.5 d
4.6 a. 3 b. 5
c. 4 d. 2
e. 1 f. 6

SELF TEST 4

4.01 family
4.02 God
4.03 teach me to take care of myself
4.04 look, think
4.05 a. food, air, water
4.06 Any order:
a. grains (whole grains)
b. vegetables
c. fruits
d. dairy (milk products, including yogurt, cottage cheese, cheese, etc.)
e. proteins
4.07 MyPlate
4.08 fats, salt, and sweets or sugar
4.09 a. lots of sleep
b. good food
c. exercise, keep clean
4.010 Teacher check
4.011 10
4.012 wool
4.013 balloon or soon
4.014 preach or eat
4.015 bread
4.016 eat or preach
4.017 soon or balloon

LIFEPAC TEST

1. Bones
2. Muscles
3. organs
4. Skin
5. five
6. look
7. think
8. God
9. exercise
10. air, food, water
11. family
12.-14. Answers will vary.
12. lots of sleep
13. exercise
14. keep clean
15.-18. Any order, any four:
15. vegetables
16. fruit
17. grains (whole grains)
18. dairy (milk products - yogurt, cottage cheese, cheese, etc.) or proteins
19. solid fats, salt, and sweets or sugar
20. ea
21. oo
22. ea
23. oo
24. ea

ALTERNATE LIFEPAC TEST

1.-5. Any order:
1. vegetables
2. fruits
3. grains (whole grains)
4. dairy (milk products)
5. protein
6. yes
7. yes
8. no
9. yes
10. no
11. yes
12. yes
13. no
14. yes
15. yes
16. yes
17. lots of sleep
18. exercise
19. keeping clean
20. ea
21. oo
22. ea
23. oo
24. ea

SCIENCE 204

ALTERNATE LIFEPAC TEST

Name ______________________

Date ______________________

Each answer = 1 point

Name the five food groups that you should eat most often.

1. ______________________
2. ______________________
3. ______________________
4. ______________________
5. ______________________

Write *yes* or *no*.

6. __________ Bones protect some of the soft parts of your body.
7. __________ Muscles help you move.
8. __________ Machines inside your body are called skin.
9. __________ Skin has small holes in it.
10. __________ We have seven food groups.

11. ________ God makes people look different.

12. ________ God lets people think differently.

13. ________ To keep in good shape, you should sit a lot.

14. ________ You need air, food, and water to live.

15. ________ Every person is made by God.

16. ________ Your are part of a family.

Name three good health habits.

17. ____

18. ____

19. ____

Write the letters to make the correct vowel sounds.

20. br ___ ___ d

21. sp ___ ___ n

22. p ___ ___ nut

23. c ___ ___ kie

24. r ___ ___ dy

SCIENCE 205

Unit 5: Pet and Plant Care

TEACHER NOTES

MATERIALS NEEDED FOR LIFEPAC	
Required	
• paper cup • dirt • grass seed • felt pen	• bottle-shaped gourd • bucket of water • beans or small stones • tape paint

ADDITIONAL LEARNING ACTIVITIES

Section 1: All About Pets

1. Bring a pet to school (fish, gerbil, snake). Hold it while you introduce this section. "I have chosen a pet for our classroom. As you can see, I chose this (fish, gerbil). His name is (pet's name). I did not choose a horse, a cow, or a rattlesnake. Why?"
   a. "We will have to take care of our new pet. Do you have any ideas about what we will need to do for (name)?"
   b. List all suggestions children offer. Assign a child for each task necessary: cleaning cage, feeding, and so forth.
2. Start a discussion about pets: "Some of you have pets at home. Can you tell us how you take care of them?"
3. Assign the students to choose an unusual pet and follow these directions: Read about it. Ask your parents to help you find information about it. Write out a list of ways this special pet must be cared for. Include food, pet's home, exercise, cleaning, and so forth.
4. As a class, read *Gator, Gator, Second Grader: Classroom Pet or Not?* by Conrad Storad. Discuss which animals make good pets and which ones do not.
5. Individually or as a class, have the students fold a paper in half. On one half, list ten or more different animals that could be pets. Find out what each animal eats. On the other half of the paper, list the kinds of foods the animals like to eat.

Section 2: All About Plants

1. Bring two plants to school. They should be the same kind of plant. One should be healthy; the other should be wilted. Tell students, "I have two plants. They are the same kind of plant. Do they both look the same? Why? What could we do to help plant number two? Why do you think plants need to be taken care of?"
2. Take a field trip to a florist shop or invite the owner to visit the class. Have the visitor show several kinds of plants and flowers and talk about where the plants grow and what kind of care they require.
3. Assign students to try growing a new plant by following these directions: Ask a parent to help you get a cutting from a house plant. Try starting it in water or dirt, depending on the plant. Record the day you started to grow it. Make a drawing of what happened each day. Record what you did each day to care for it. Report what happened after two weeks.

ANSWER KEYS

SECTION 1

1.1 Teacher check
1.2 Mother
1.3 Lora
1.4 Mother
1.5 Lora and Keith
1.6 pecking, scolding, swinging
1.7 things or pets
1.8 pretty
1.9 different
1.10 glass
1.11 water, seed
1.12 friend
1.13 shapes
1.14 home
1.15 Teacher check
1.16 first, last, last, first
first, first, first, first
1.17 chin, cheese, chop
1.18 chick
1.19 castle, chest
1.20 horse
1.21 snake
1.22 pet
1.23 shells, shoes, sheep, chain, thin, thick
1.24 first, first, last,last
1.25 thumb
1.26 Teacher check
1.27 Lora wanted to ride a horse.
1.28 Food for a horse costs too much money. (or)
The yard was too small.
1.29 Teacher check
1.30 wh, th, wh, th, wh, th
1.31 Teacher check
1.32 Answers will vary.
1.33 Examples:

1. cat food	2. dish—food
3. dish—water	4. basket
5. toy	6. catnip
7. blanket	8. brush
9. box of dirt	10. cat treats

1.34 noise
1.35 Gerbils
1.36 box
1.37 paper
1.38 wheel
1.39 gold
1.40 kitten
1.41 well
1.42 love
1.43 talk
1.44 Example:
Look for a pet that is small enough for home.
Look for a pet that everyone likes.
Look for a pet that is easy to care for.
1.45 Example:
have a dish of water; feed on time; have a bed; talk to your pet; love your pet; rub his head
1.46 The pet will not live.
1.47 Teacher check
1.48 wheel
1.49 chair
1.50 sheep
1.51 thimble
1.52 shell
1.53 wheat
1.54 thin

SELF TEST 1

1.01 yes
1.02 no
1.03 no
1.04 yes
1.05 no
1.06 yes
1.07 yes
1.08 no
1.09 yes
1.010 no
1.011 √
1.012 O
1.013 √
1.014 O
1.015 √
1.016 friend
1.017 mice
1.018 horses
1.019 rats
1.020 kitten
1.021 shots
1.022 list
1.023 fits
1.024 talk
1.025 water

SECTION 2

2.1 Teacher check
2.2
1. plant
2. Lora
3. ar
4. need
5. teacher
6. seeds

2.3-2.4 Teacher check
2.5 sweet potato, water
2.6 roots
2.7 shoots
2.8 one foot
2.9 Answers will vary.
2.10 Teacher check
2.11 Teacher check
2.12 Any order: leaf, stem, roots, fruit, seed, flowers
2.13 Examples: carrot, tomato, banana, apple, celery, potato
2.14 eat
2.15 breathe
2.16 breathe
2.17 plants
2.18 plant or gourd
2.19 Any order: leaves, stem, roots, fruit, flower, seeds
2.20 plants
2.21 job
2.22 care
2.23 Teacher check
2.24 arm, car, farm, jar, star, park
2.25 yes
2.26 no
2.27 no
2.28 yes
2.29 yes
2.30 corn
2.31 cord
2.32 fork
2.33 north or storm
2.34 fort
2.35-2.38 Any order:
2.35 car
2.36 jar
2.37 tar
2.38 star
2.39 germ, hurt, chirp, burn, bird, first
2.40 hurt
2.41 bird
2.42 chirp
2.43 first
2.44 burn
2.45 cactus
2.46 water
2.47 live
2.48 Examples:
Water them just right.
Keep them near sunlight.
Don't let them get too hot or too cold.

SELF TEST 2

2.01 plants
2.02 friend
2.03 water
2.04 grow
2.05 cactus
2.06 care
2.07 big
2.08 job
2.09 horse
2.010 flowers
2.011 O
2.012 √
2.013 O
2.014 O
2.015 √
2.016 √
2.017 √
2.018 O
2.019 √
2.020 O
2.021 shapes
2.022 snake
2.023 gold
2.024 kitten
2.025 love
2.026 water
2.027 care
2.028 shoe
2.029 inside
2.030 breathe

LIFEPAC TEST

1. √
2. O
3. √
4. O
5. √
6. O
7. √
8. √
9. O
10. yes
11. yes
12. no
13. yes
14. no
15. no
16. yes
17. yes
18. yes
19. yes
20. no
21. toy
22. rats
23. care
24. grow
25. inside
26. vine
27. roots
28. love
29. The pet would not live.
30. The plant would not live.
31. Answers will vary.

ALTERNATE LIFEPAC TEST

1. yes
2. no
3. no
4. yes
5. yes
6. no
7. yes
8. yes
9. yes
10. a fish
11. An apple tree
12. friend
13. cactus
14. dog
15. sweet potato
16. horse
17. home
18. soft
19. care
20. inside
21. plant
22. pet
23. plant
24. pet
25. pet

SCIENCE 205

ALTERNATE LIFEPAC TEST

Name ____________________

Date ____________________

Each answer = 1 point

Write *yes* or *no*.

1. ________ Plants need water.
2. ________ Pets need a bike.
3. ________ Plants need to walk around.
4. ________ Pets need love.
5. ________ Plants help the air you breathe.
6. ________ Plants need a chair.
7. ________ Pets need water.
8. ________ Plants need the sun.
9. ________ Pets need food.

Circle and write the correct answer.

10. A good pet for you would be ____________________ .

an elephant a fish

11. ______________________ should be planted outside.

An apple tree — A house plant

12. A pet is your ______________________ .

toy — friend

13. A ______________________ needs very little water.

cactus — tree

14. Jim plays with his ______________________ .

plant — dog

15. A ______________________ plant will grow in water.

tree — sweet potato

16. You can ride a ______________________ .

horse — snake

17. A pet is an animal you keep at ______________________ .

home — the zoo

18. Kittens are ______________________ .

hard — soft

19. Plants need your ______________________ .

car — care

20. House plants grow ______________________ .

inside — outside

Write *pet* or *plant* next to each word.

21. __________ sweet potato

22. __________ dog

23. __________ grass

24. __________ fish

25. __________ cat

SCIENCE 206

Unit 6: Your Five Senses

TEACHER NOTES

MATERIALS NEEDED FOR LIFEPAC

Required

- mirror
- microscope
- magnifying glass
- telescope
- ball-point pen
- 1 cup butter or margarine
- 2 1/2 cups flour
- 1 teaspoon vanilla
- 1/2 cup confectioner's sugar
- 1/4 teaspoon salt
- 3/4 cup chopped walnuts
- microwave popcorn

ADDITIONAL LEARNING ACTIVITIES

Section 1: You Can See

1. Present a model or a detailed poster of an eye. Discuss the following questions.
   a. What can you do with your eyes?
   b. How many parts of the eye can you name?

   Introduce *eyebrow*, *eyelash*, *eyelid*, and *eyeball* as vocabulary words. Discuss the function of each part.
2. Discuss different reasons why some people can't see very well. Explain that eyeglasses can often correct problems with our vision. Bring in a pair of non-prescription reading glasses and allow each student to try them to see the effect. Allow any students who do wear glasses talk about how glasses help them to see better.
3. Present a microscope, telescope, and a magnifying glass. Demonstrate how there items are used. Let students take turns using them.
4. Discuss what it means to be blind. Ask the students, "What kinds of problems might a blind person face?" Discuss crossing streets, using white canes, guide dogs, and Braille.
5. Take turns blindfolding the students, one at a time. Ask them how they feel about not being able to see. Discuss how the other senses (hearing, smell, touch, and taste) become more important to a person who has lost their sight and how a blind person learns to rely on these other senses in everyday life.
6. Invite a blind person to come and speak to the class.
7. Invite the school nurse to come and explain how the eye test is used. Have her test the students.
8. Allow each student to make their own drawing of the eye and label all the parts. Allow students to share their drawings and explain how the eye and its parts work.
9. Have the students use scraps of material, string, macaroni, cereal, or sand to make a "touch" picture that a blind person could enjoy.
10. Take the students on a field trip to a blind school in your area and allow them to see and experience first-hand how blind children learn and play together.

Section 2: You Can Smell

1. Discuss with the students other functions of the nose, such as breathing.
2. Allow students to choose a place (grocery store, circus, zoo, woods, ocean, shopping mall, etc.) and tell or write about the kinds of smells they might experience there.
3. Go on a smelling walk! Take a walk around your school or neighborhood and list all the smells experienced.
4. Collect several items that have distinctive smells such as: lemon, banana, mint, orange peel, pine needles, vanilla, ginger, cedar wood, chocolate, garlic, onion, moth balls, perfume-soaked cotton, coffee, pencil shavings, etc. Keep the items separated and enclosed in plastic containers so that the smells do not mix. Blindfold students (or have them close their eyes) and allow them to smell one or several of the items. Ask these questions:
   a. What do you smell?
   b. Is the odor strong, faint, pleasant, or unpleasant?
   c. What do you think of when you smell this scent? Why?
5. Collect two of each item such as was listed in the previous activity. Place in separate containers with holes punched in the top. See if students can correctly match the pairs by smelling.

Section 3: You Can Hear

1. Explain to the class that deaf people cannot hear. Because of this, their other senses become more important to them. Talk about the ways a deaf person would use their other senses to adapt.
2. Explain to the class that deaf people communicate in different ways (reading lips and sign language). Talk about some common "signs" we all use to communicate each day (examples: hi, bye, stop, shh, yes, no, etc.).
3. Review the sign language letters of the alphabet and teach each student how to "sign" their name. You might also obtain a good sign language book and teach your class some simple signs. Or, learn how to sign a song, such as "Jesus Loves Me."
4. Play "guess the sound" game. Use recorded sounds, or the teacher or students may make actual sounds.
5. Make a drawing of the ear on a poster or bulletin board display. Label the parts. Explain to the class the way the ear works.
6. Have the students write or tell about their favorite sounds. Children may make pictures to accompany their stories.
7. If possible, visit a deaf school and allow the children to see and experience first hand how deaf children learn, play, and interact with each other. Follow up with a class discussion.

Section 4: You Can Taste

1. Make a large drawing of the tongue and divide it into the four different areas of taste: sweet, sour, salty, and bitter. Bring in one example of each type of food and allow the students to taste. Then, write the name of that food or paste a picture of it onto the correct

section of the tongue drawing. (You may expand this exercise by asking the students to name different foods and then determine which portion of the tongue would taste it.)

2. Explain to the class that the nose (the sense of smell) is responsible in part for the flavor of food. To demonstrate this, blindfold (or have each student cover their eyes) and have the students also hold their nose. Then, allow them to taste something and see if they can distinguish the flavor. Note: different flavored jelly beans work well for this test since they all have the same texture.
3. Cut out magazine pictures of foods. See if the students can correctly categorize them as sweet, salty, sour, or bitter.
4. Have each student write or tell about their favorite food and describe how it tastes and why it is a favorite.

Section 5: You Can Feel

1. Review the first four senses that have been studied. Ask the students, "What do you think the fifth sense might be?" Allow the students to feel several objects and describe what they feel. Record the "feel" words on the board.
2. Talk about the importance of the sense of touch for someone who cannot see. Have students close their eyes and feel objects from their desks. Allow them to describe how it feels and try to guess what it is.
3. Bring in various grades of sandpaper and see if the students can tell the difference through touch.
4. Write various "feel" words on the board and see how many examples the students can give of each (examples: smooth, rough, cold, warm, soft, hard, etc.).
5. Place some familiar objects (cup, spoon, ball, block, sponge, rock, cotton ball, leaf, pine cone, feather, coin, etc.) into a box or bag. Have students (one at a time) close their eyes and pull out an object. See if they can identify through touch.
6. Collect pairs of objects (2 identical bottle caps, paper clips, marbles, etc.) and place one item from each pair into a box. Place the other item from the pair into a different box. See if students can reach into each box (with eyes closed) and correctly find a matching pair of items.
7. Have the students create touch boards by following these directions: Divide a large piece into sections. Glue on popsicle sticks to divide the cardboard into sections. In each section, glue objects that have different textures. For example, sand, small pebbles, string, feathers, etc. could all be glued into different sections. Everyone's board should be a little different. Allow students to compare the different textures. (Note: Materials can be brought in from home and/or could be gathered on a nature walk.)
8. Bring in some simple children's wooden puzzles containing about 4 to 6 pieces each. See if students can work the puzzles while blindfolded or with eyes closed (using only the sense of touch).
9. If a microwave oven is not available for the final experiment in the unit, modify the instructions and checklist to accommodate whatever method you have for popping corn.

Tying It All Together

Try the following ideas as a means of demonstrating how all our senses work together.

1. Make a "sense chart" for the class. Draw or make a chart with a picture of an eye, ear, nose, tongue, and hand. Write down words or hold up pictures that "go" with each of the senses that are represented. Examples might include:

   eye: light bulb, sun, flashlight

   tongue: ice cream, hamburger

   ear: musical instruments, radio

   hand: pencil, pen, glove

   nose: garbage, skunk, bread, apple pie

   Explain that many of the items could go under several categories therefore demonstrating how our senses work together to tell us about the outside world.

2. Have a food party. Bake cookies or prepare other types of food together. Together, make a list or talk about what each of the five senses experienced. Prompt students to describe the tastes and texture of the food.
3. Create sensory stations around the classroom. Set up tables with hands-on materials for kids to touch, smell, taste, hear, and see. Use caution in selecting items that my cause an allergic reaction. Some possible station items might include:

   hearing: bells, drums, whistles, rice-filled containers

   touch: sandpaper, play dough, ice, finger paint, rocks

   vision: microscope, magnifying glass, colored water, prism

   smell: vanilla, perfume, chocolate, spices

   taste: jelly beans, cookies, crackers, fruit slices, salty chips, dill pickles

4. Go on a sensory nature walk. Allow the students to collect several items apiece. Then, share what was collected and allow students to describe their objects' sensory characteristics.
5. Brainstorm as a class and see if anyone can name one object that appeals to every one of the five senses. Clue: Are there any noisy edibles? Are there any tasty sounds?
6. Obtain a teacher's collection of age-appropriate books from your local public library on the topic of the five senses. Have books available in a special "reading corner" for use during free time.
7. Take a field trip to a children's interactive science museum that includes displays about the five senses and how the senses work individually and together.

Administer the LIFEPAC Test.

The test is to be administered in one session. Give no help except with directions.
Evaluate the tests and review areas where the students have done poorly.
Review the pages and activities that stress the concepts tested.
If necessary, administer the Alternate LIFEPAC test.

ANSWER KEYS

SECTION 1

1.1 Answers will vary.
1.2 Answers will vary.
1.3 Answers will vary.
1.4 Drawings will vary.
1.5 eyelash — keeps out dirty things
eyeball — does the seeing
eyebrow — makes a wall
eyelid — opens and closes
1.6 a. eyebrow
b. eyelid
c. eyelash
d. eyeball
1.7 north corn short
thorn fork storm
porch horn born
Teacher check
1.8 up close
1.9 nurse
1.10 glasses
1.11 letters
1.12 "...and every eye shall see him."
Revelation 1:7
1.13 microscope — to see small things
magnifying glass — to make words bigger
telescope — to see stars
1.14 Teacher check
1.15 Teacher check
1.16 love one another
1.17 yes
1.18 yes
1.19 yes
1.20 yes
1.21 no
1.22 yes
1.23 no
1.24 yes
1.25 no
1.26 yes

SELF TEST 1

1.01 eyes
1.02 clean
1.03 closed
1.04 safe
1.05 bad
1.06 Braille.
1.07 God's world.
1.08 opens and shuts.
1.09 read the letters.
1.010 makes things look bigger.

SECTION 2

2.1 nostrils
2.2 air
2.3 air
2.4 smell
2.5 Answers will vary.
2.6 Answers will vary.
2.7 Teacher check, Examples: smelling gas from a gas stove or smelling smoke.
2.8

apples	in the kitchen
flowers	at the bakery
hay	at the market
supper	at the farm
cake	in the garden

2.9 Teacher check

SELF TEST 2

2.01 nose
2.02 eyes
2.03 holes
2.04 safe
2.05 parts
2.06 Braille
2.07 Teacher check
2.08 Teacher check

SECTION 3

3.1 ears
3.2 cup
3.3 soft
3.4 hurt
3.5 three
3.6 God loves you
3.7 Teacher check
3.8 Answers will vary.
3.9 Answers will vary.
3.10 Teacher check

SELF TEST 3

3.01 with our sense of hearing.
3.02 with your hands.
3.03 hurt our sense of hearing.
3.04 hearing, and hearing by the Word of God.
3.05 seeing, smelling
3.06 seeing, smelling
3.07 seeing, hearing
3.08 seeing, smelling, hearing

SECTION 4

4.1
1. bitter
2. sour
3. salt
4. sweet

4.2
3 cracker
2 dill pickle
4 candy
3 popcorn
2 lemon
4 cake
4 jelly
3 pretzel

4.3 no
4.4 yes
4.5 no
4.6 no
4.7 yes
4.8 Teacher check

SELF TEST 4

4.01

See
rain
sun
cupcake
whistle
ice
fire
smoke
plane
cake
paint
traffic
apple
cookie
lemonade
phone
fish

Hear
rain
song
whistle
laugh
fire
whisper
plane
noise
traffic
phone

Smell
rain
cupcake
fire
smoke
cake
paint
apple
cookie
lemonade
fish

Taste
cupcake
ice
cake
apple
cookie
lemonade
fish

SECTION 5

5.1 Teacher check
5.2 Any order:
a. the tip of your tongue
b. the tip of your nose
c. the tips of your fingers
5.3 Examples:

a. seeing	a sunset
b. smelling	a flower
c. hearing	music
d. tasting	a cookie
e. feeling	a soft blanket

5.4 yes
yes
yes
yes
yes
Teacher check

SELF TEST 5

5.01 your eyes
5.02 with your hands
5.03 using Braille
5.04 on your tongue
5.05 in different places
5.06 a flower
5.07 touch
5.08 on most places
5.09 tips of our fingers
5.010 three
5.011-5.015: Senses may be listed in any order.
Examples:

seeing	houses
smelling	smoke
hearing	birds singing
tasting	ice cream
feeling	sandpaper

LIFEPAC TEST

1.-5. Examples:

1.	hearing	thunder
2.	seeing	clouds
3.	smelling	apple pie baking
4.	feeling	a soft kitten
5.	tasting	pizza

6. yes
7. yes
8. yes
9. no
10. no
11. a "seeing-eye dog."
12. hearing the Word of God.
13. on your tongue.
14. in different places.
15. reads Braille.
16. pointed things near your eyes.
17. makes letters look bigger.
18. good and bad things.
19. listen
20. tip
21. nose
22. hear
23. blind

ALTERNATE LIFEPAC TEST

1. no
2. yes
3. no
4. yes
5. no
6. clean
7. white
8. nostrils
9. safe
10. hands
11. loud
12. Popcorn
13. four
14. touch
15. most places
16.-20. Senses may be listed in any order.
Examples:

sight	a car
smell	bread baking
hear	music
taste	pizza
feel	a soft blanket

SCIENCE 206

ALTERNATE LIFEPAC TEST

Name ____________________

Date ____________________

Each answer = 1 point

Write *yes* or *no* in front of each sentence.

1. ________ You smell with your feet.

2. ________ You can see the sky with your eyes.

3. ________ An ice cube feels warm when you touch it.

4. ________ We hear music and many other sounds with our ears.

5. ________ Ice cream has a bitter taste.

Circle and write the correct answer.

6. Tears help keep the eyes ____________________ .

clean dry

7. Blind people use ____________________ canes.

black white

8. A nose has two holes called ____________________ .

nostrils nobs

9. Smells help you to be ______________ .

safe soft

10. You can talk to deaf people with your ________________ .

feet hands

11. Sounds that are too _______________ can hurt your sense of hearing.

loud soft

12. _________________ has a salty taste.

Candy Popcorn

13. Your tongue can taste __________ different kinds of taste.

two four

14. People brought children to Jesus so that He could ______________ them.

touch see

15. The sense of touch is located on _______________________ of your body.

most places the inside

Name the five senses. Give an example of each.

16. _________________________ _________________________

17. _________________________ _________________________

18. _________________________ _________________________

19. _________________________ _________________________

20. _________________________ _________________________

SCIENCE 207

Unit 7: Physical Properties

TEACHER NOTES

MATERIALS NEEDED FOR LIFEPAC

Required

- crayons
- plain white paper
- paste, crayons
- 3 cups of flour
- 2 cups of salt
- 1 tablespoon of powdered alum
- 2 cups of boiling water
- 3 tablespoons of oil
- food coloring
- flashlight
- CD (compact disc)
- white paper
- finger paints in the following colors: red, yellow, blue, black, and white
- paper plates (at least 6)
- old shirt or apron
- paper towels

ADDITIONAL LEARNING ACTIVITIES

Section 1: All About Colors

1. Ask students to name all the colors they can think of. Write the names on the board.
2. Name an object. Ask students to name the best color for that object. Continue naming a variety of different types of objects.
3. Provide finger paints. Allow the children to experiment with mixing the different colors (primary as well as black and white). Then, through discussion, reinforce the concepts of mixing colors and light and dark colors.
4. Make a class quilt! Provide each student with a white square of paper with holes prepunched around the edges. Allow students to decorate their patches with various colored crayons, markers, and/or paints. "Sew" all the patches together with colorful yarn and hang in classroom.
5. For a set period of time (perhaps 10–15 minutes), have students list as many objects as they can think of and the color associated with each (example: a lemon—yellow). Or, the teacher can name a color and allow the students to list as many objects as they can that are associated with that color.
6. Have the students make a color chart: Use primary colors to make new colors. Write on the chart the colors they mixed. Example: yellow + red = orange
7. Try this experiment to demonstrate that mixing many colors makes black. You will need these items: scissors, white paper coffee filter, black marker (not permanent), water, a coffee mug. Follow these steps:
   a. Cut a circle out of the coffee filter.
   b. With the black marker, draw a line across the circle, about 1 inch up from the bottom.
   c. Put some water in the mug—enough to cover the bottom.

   d. Curl the filter paper circle so that it fits inside the cup. (Make sure the bottom of the circle is in the water.)

   e. Watch as the water flows up the paper. When it touches the black line, you'll start to see some different colors.

   f. Leave the paper in the water until the colors go all the way to the top edge. How many colors can be seen?

   This experiment may also be done by drawing a black spot in the middle of the coffee filter. Place the coffee filter circle on a saucer and place a few drops of water on the spot. In a few minutes you'll see rings of color that go out from the center to the edges.

8. Ask the children if they have ever seen a rainbow. Under what conditions did they see it? Discuss what causes rainbows.

9. Tape a large, long sheet of paper to the wall in your classroom and allow the children to paint or color a mural depicting the Bible story of Noah's ark.

10. Some students may attend a liturgical church. If so, you may discuss the different colors associated with the liturgical seasons in the church year.
    Examples: Lent—purple, Easter—white, Pentecost—red, Ordinary time—green, Advent—purple.

    You may also wish to invite a minister or other church member to speak with the class about the use of colors throughout the church year and what they symbolize.

Section 2: Shapes and Sizes

1. Introduce the vocabulary words rectangle, square, circle, triangle, and octagon and the definitions of each.
2. Prepare construction cut-outs of various shapes (squares, circles, triangle, rectangle, and octagon). Make the shapes different sizes and colors. Present the shapes one at a time and have the students name the shape and describe it.
3. Introduce irregular shapes. You might explain, "Everything has a shape. Dogs, people, trains, even lakes have a certain shape. We call these shapes irregular." Allow students to draw some irregular shapes on the board.
4. Give students a list of objects. Have them write shape and size words for the objects.
5. Ask students to observe the different types of highway and traffic signs they see when traveling. Then, have them make signs. Allow them to show their signs to the class and describe their shapes, sizes, and purposes.
6. Provide each student with a shape stencil (one which has a square, triangle, rectangle, etc.). Have each student trace and then color their own shape picture. Pictures may be displayed on a class bulletin board about shapes and sizes.
7. Have the students use old magazines to find pictures that contain squares, rectangles, triangles, circles, etc. Monitor as they cut out and paste the pictures on pages of plain white paper, and then write the name of the shape and a size word under each picture. Display pages on a bulletin board or compile into a class scrapbook about shapes and sizes.

8. Read the book *Shapes, Shapes, Shapes* by Tana Hoban aloud to the class. Then discuss the following questions:
   a. What shapes can you see in our classroom?
   b. Are there more shapes in your kitchen or bathroom?
   c. What shapes are found in your room at home?
   d. What shapes are you wearing today?
9. Go on a shape walk! Take a walk around your neighborhood. Take pictures of the various shapes you find. Print out the pictures and display on a bulletin board or in a class scrap-book. Write descriptive words under each shape picture. You may discuss these questions:
   a. What shapes are most common?
   b. Where are squares, circles, etc. found in nature?
   c. Which is easier, finding shapes in natural- or human-made structures?
10. Try this activity. Tell the students: Imagine a place that is made out of only one shape. Draw a picture and write or tell about this place.

Section 3: How Things Feel

1. Explain to the class: "All objects feel a certain way. They can feel hot or cold, wet or dry, smooth or rough, hard or soft." Pass around several items of various textures to the class. Allow the students to describe how these objects feel. Then discuss: How would a kitten's fur feel? a marshmallow? a rock?
2. Write a "feel" word on the board. (Examples: hard, soft, wet, dry, rough, smooth, etc.). See how many objects the students can name which would fit the word written on the board.
3. Play the simile game. The teacher or student says a common simile leaving off the end of the sentence. Other students must try to guess the answer.

   Examples: "As blue as ______________." (Answer: the sky)

   "As rough as ______________." (Answer: sandpaper)
4. Provide a box or bag filled with objects of various sizes, shapes, and textures. Have students close their eyes and pick one object. Then have them describe their object according to its color, shape, size, temperature, and texture.

Administer the LIFEPAC Test.

The test is to be administered in one session. Give no help except with directions.
Evaluate the tests and review areas where the students have done poorly.
Review the pages and activities that stress the concepts tested.
If necessary, administer the Alternate LIFEPAC test

ANSWER KEYS

SECTION 1

1.1 orange
1.2 purple
1.3 green
1.4 darker
1.5 lighter
1.6 gray
1.7 Any order:
a. red
b. yellow
c. blue
1.8 a. orange
b. purple
c. green
d. light blue
e. pink
f. gray
1.9 Examples: green, blue
1.10 Examples: red, orange, brown
1.11 a cherry
1.12 the sky
1.13 the grass
1.14 the sun
1.15 the night
1.16 a tree trunk
1.17 yes
1.18 The bands of color shift and move.
1.19 Rainbow colors are reflected.
1.20 Teacher check, Example: The bands of color become wider or thinner.
1.21 yellow
1.22 red
1.23 brown
1.24 rainbow
1.25 Roy G. Biv
1.26 Circled colors: orange, red, violet, green, blue, indigo, yellow
1.27 a and c
1.28 Teacher check
1.29 Teacher check
1.30 red
1.31 orange
1.32 red, white, and blue
1.33 white and yellow
1.34 green
1.35 red and green

SELF TEST 1

1.01 Any order:
a. red
b. yellow
c. blue
1.02 a. red
b. orange
c. yellow
d. green
e. blue
f. indigo
g. violet
1.03 orange
1.04 light blue
1.05 green
1.06 white
1.07 gray
1.08 b and d
1.09 orange, brown
1.010 red, white, blue
1.011 red
1.012 green
1.013 yellow, white
1.014 red, green

SECTION 2

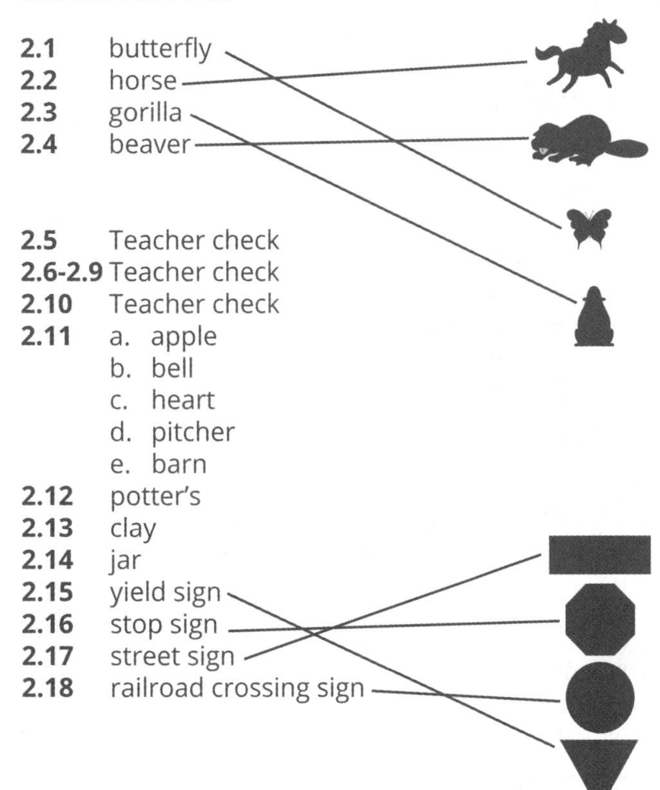

2.1 butterfly
2.2 horse
2.3 gorilla
2.4 beaver

2.5 Teacher check
2.6-2.9 Teacher check
2.10 Teacher check
2.11 a. apple
b. bell
c. heart
d. pitcher
e. barn
2.12 potter's
2.13 clay
2.14 jar
2.15 yield sign
2.16 stop sign
2.17 street sign
2.18 railroad crossing sign

SELF TEST 2

2.01 yes
2.02 yes
2.03 no
2.04 yes
2.05 no
2.06 yes
2.07 no
2.08 no
2.09 jar
2.010 yield
2.011 blue
2.012 stop
2.013 Either order: shapes, sizes
2.014 Either order: white, yellow
2.015

Across	Down
4. six	1. basket
5. potter	2. square
6. down	3. rectangle
8. green	7. red
9. circle	

SECTION 3

3.1 Teacher check
3.2 The wool of a sheep is fuzzy.
3.3 Sandpaper is rough.
3.4 A desk is smooth.
3.5 Mud is hard.
3.6 Pears are soft.
3.7 Glass is smooth.
3.8 Carpet is soft.
3.9 A sweater is hard.
3.10 Oatmeal is soft.
3.11 A muffin is hard.
3.12-3.20 Teacher check
3.21 H
3.22 S, S
3.23 S, S, S
3.24 H, S, H
3.25 H, S
3.26 hard
3.27 soft
3.28 soft
3.29 hard
3.30 soft
3.31 hard
3.32

		d	u	s	v	r	m			
	b	d	q	s	t	h	i	c	k	
x	r	o	u	g	h	m	l	p	v	
z	n	o	p	z	f	c	d	e	f	d
r	t	h	i	n	n	o	s	s	u	f
v	s	t	u	r	w	o	t	o	z	d
l	m	w	v	z	x	l	f	f	z	d
s	o	a	n	h	f	v	e	t	y	e
c	d	r	b	d	h	i	k	j	k	
	s	m	s	m	o	o	t	h		

SELF TEST 3

3.01 red, yellow, blue
3.02 pink
3.03 warm color
3.04 three sides
3.05 spring
3.06 orange
3.07 warm
3.08 wet
3.09 hard
3.010 fuzzy
3.011 soft
3.012 smooth
3.013 rough
3.014 hard
3.015 hard
3.016 smooth
3.017 Roy G. Biv
3.018 a. red
b. orange
c. yellow
d. green
e. blue
f. indigo
g. violet

LIFEPAC TEST

1. primary colors
2. rectangle
3. orange leaves
4. triangle
5. red
6. green
7. brown
8. red
9. orange
10. white
11. blue
12. yield sign
13. stop sign
14. street sign
15. railroad crossing sign
16. Ben found a big cob of corn.
17. Sam went to the car race.
18. Our cat does not like mice.
19. rough
20. hard
21. smooth
22. soft
23. circle
24. rectangle
25. square
26. triangle
27. octagon

ALTERNATE LIFEPAC TEST

1. darker
2. primary
3. lighter
4. rainbow
5. triangle
6. square
7. rectangle
8. potter
9. Signs
10. fleece
11. stop
12. red and green
13. white and yellow
14. rough
15. smooth
16. round and flat shape
17. yield
18. b, d
19. Roy G. Biv

SCIENCE 207

ALTERNATE LIFEPAC TEST

Name ______________________

Date ______________________

My Score

16/20

Each answer = 1 point

Circle the correct answer.

1. Mixing black with another color makes the other color ______ .

lighter darker

2. Red, yellow, and blue are called ______ colors.

primary basic

3. Mixing white with another color makes the other color ______ .

lighter darker

4. The _____ is a very special thing that God made with many colors.

cloud rainbow

5. A shape with three sides and three points is a ______ .

square triangle

6. A ______ is the same on all four sides and has four corners.

square rectangle

7. A ______ has four corners, two long sides, and two short sides.

square rectangle

8. Jeremiah watched the ______ make a jar on his wheel.

potter carpenter

9. ______ are used to let us know something or to give us directions.

Stones Signs

10. Gideon's sign was in the feel of a ______ and the feel of the ground.

rug fleece

Draw lines to match.

11.	red sign with eight sides ▸	◂	rough
12.	Christmas colors ▸	◂	white and yellow
13.	Easter colors ▸	◂	stop
14.	sandpaper ▸	◂	smooth
15.	paper ▸	◂	red and green
16.	circle ▸	◂	yield
17.	triangle-shaped sign ▸	◂	round and flat shape

Answer these questions.

18. Read the sentences. Two of the sentences tell what God promised Noah the rainbow would show. Circle the letters of the two correct sentences.

a. The rainbow would lead to a pot of gold.

b. The rainbow would be a sign of God's promise not to cause another great Flood over the whole earth.

c. The rainbow would help keep the clouds in the sky.

d. The rainbow would show God's love toward all people and creatures.

19. What is the name of the person that helps you remember the colors of the rainbow?

____

SCIENCE 208

Unit 8: Our Neighborhood

TEACHER NOTES

MATERIALS NEEDED FOR LIFEPAC	
Required	Suggested
• magazine • scissors	• Bible, King James version

ADDITIONAL LEARNING ACTIVITIES

Section 1: What Is Environment?

1. Define *environment* as *everything that is around you*. Ask students to name several things in their environment. List student responses in two columns on the board. In one column list *things living*, in the other *things not living*.
2. Discuss with the students the importance of water, air, sun, and soil to the environment. Stress that these are treasures given by God to help support living things.
3. Have students take turns explaining a food chain in the sea, from Section 1.
4. Assign the students to draw a picture of a food chain for an apple tree or a tomato plant. Write a few sentences to explain.
5. Individually or as a class, have the students fold a paper in half, lengthwise. List all the living things in your home environment on one side. List all the nonliving things in your home environment on the other half.

Section 2: What Is Pollution?

1. Ask students to observe while you scatter some trash on the floor and make loud noises. (Bang a pan, or play loud music.) Explain that trash and noise are two kinds of pollution. Anything that makes water, air, or the ground dirty is pollution. Ask students to think of other kinds of pollution. Discuss smog and bad odors.
2. Ask students to think of ways to avoid pollution. Introduce the vocabulary word *recycling*. Tell students, "One way to avoid pollution is to recycle old cans, bottles, and papers instead of throwing them in the trash." Discuss recycling centers in your area. Encourage children to ask parents to help in recycling efforts.
3. Visit a recycling center in your area. Have students bring in newspapers, cans, or bottles (whichever applies to your area) to the center.
4. Have students draw pictures of pollution. Have them take turns showing their pictures.
5. Have students write a story about something someone did to stop pollution.

Section 3: What Is Ecology?

1. Discuss these questions with the students:
   a. What kinds of things might an ecologist do?
   b. How does fertilizer help the home environment?
   c. If a logger cuts down a tree, what could he do to renew the environment?
   d. Can you name something you do to keep your environment clean and healthy?
2. Have the students use old magazines to find and cut out pictures of pollution. Direct them to paste the pictures on one side of a plain piece of paper and use the other side to draw pictures of the way they would like that environment to look.
3. Assign the students to make a list of three things they can do to make their environment clean and healthy. Have them write the days of the week across the top of a piece of paper. They should check themselves every day for a week to see if they are personally concerned about ecology.

Administer the LIFEPAC Test.

The test is to be administered in one session. Give no help except with directions.
Evaluate the tests and review areas where the students have done poorly.
Review the pages and activities that stress the concepts tested.
If necessary, administer the Alternate LIFEPAC test.

ANSWER KEYS

SECTION 1

1.1 Teacher check
1.2 children, rabbit, carrots
1.3 yes
1.4 no
1.5 yes
1.6 no
1.7 yes
1.8 yes
1.9 no
1.10 yes
1.11 Teacher check
1.12 storehouse
1.13 charge, treasures
1.14 environment
1.15 food
1.16 water
1.17 drink, cook, bathe
1.18 Teacher check
1.19 sh, ch, th
ch, sh, th, sh
th, ch, sh, ch
th, sh, ch, th
1.20 Examples: water, soil, air
1.21 ✓ Soil is not a part of our environment.
1.22 All living things are part of our environment.
1.23 ✓ Plants and animals are not living things.
1.24 Plants need our care.
1.25 ✓ Only living things are part of our environment.
1.26 ✓ Only some living things take something out of the environment.
1.27
1. cycle
2. environment
3. treasures
4. storehouse
5. earth

1.28 cedar trees
1.29 Lebanon
1.30 King Solomon
1.31 They became bare.
1.32 tree — living
air — not living
fish — living
water — not living
soil — not living
house — not living
1.33 breathe
1.34 storehouse of treasures
1.35 Lebanon
1.36 clean environment
1.37 drink
1.38 around you
1.39 more
1.40 less
1.41 Drawings will vary.
1.42 Answers will vary.

SELF TEST 1

1.01 no
1.02 yes
1.03 yes
1.04 no
1.05 yes
1.06 no
1.07 yes
1.08 yes
1.09 yes
1.010 no
1.011 Answers will vary.
Examples: plants, animals, flowers, rocks, soil
1.012 trees
1.013 bare
1.014 place
1.015 feed
1.016 number
1.017 clean
1.018 care
1.019 eat
1.020 Example:
Everything around me is my environment.

SECTION 2

2.1 taking, shopping
2.2 looking, shopping
2.3 piling
2.4 going, building
2.5 Mother
2.6 Lynn
2.7 Mother
2.8 Ron
2.9 Saturday
2.10 shopping center
2.11 busy street
2.12 up
2.13 pollution (dirty)
2.14 elevator
2.15 smog
2.16 fog
2.17 packages
2.18 ✓ smog
2.19 bank
2.20 ✓ pollution
2.21 planes
2.22 Smog is smoke and fog.
2.23 Cars burn fuel.
2.24 Smog makes some people sick.
2.25

window	owe	own
grown	fellow	bellow
blowing	glowing	shadow

2.26 fellow bellow
2.27 owe own
2.28 blowing glowing
2.29 owe own
2.30 Teacher check
2.31 smoke and fog
2.32 repeats
2.33 where you live
2.34 clean environment
2.35 to use again
2.36

cow	mountain
cloud	flower
owl	town
shout	frown

2.37 machines, dogs, cars, people
2.38 hearing
2.39 hearing
2.40 noise
2.41 power
2.42 pollution
2.43 turn down
2.44 lots of noise
2.45 passed them
2.46 hurt hearing
2.47 Teacher check

SELF TEST 2

2.01
flowers
✓ trash
✓ smog
mountains
✓ noise
✓ waste
✓ fuel
fertilizer

2.02
✓ trees
play
✓ flowers
✓ animals
cycle
✓ plants

2.03 no
2.04 yes
2.05 yes
2.06 yes
2.07 yes
2.08 no
2.09 yes
2.010 no
2.011 yes
2.012 no
2.013 around you
2.014 smoke and fog
2.015 repeats
2.016 storehouse of treasures
2.017 use again
2.018 breathe
2.019 Example: The things around me are my environment.
2.020 Example: Pollution is dirty water, air, or ground.

SECTION 3

3.1 Teacher check
3.2 Teacher check
3.3 Example:
Save old cans
Save old papers
3.4 Example:
The place around you is your environment.
3.5 Answers will vary.
3.6 Teacher check
3.7
tall
✓ lived a long time ago
hurt his hand with a saw
✓ made wonderful things
✓ loved birds, trees, mountains, and lakes
✓ lived on a farm
3.8
A MAN WHO MADE THINGS
✓ A MAN WHO LOVED GOD'S WORLD
A MAN WHO HURT HIS EYE
3.9 things
3.10 made
3.11 hurt
3.12 rule
3.13 Example:
It opened and closed books.
3.14 Example:
Because he loved birds, trees, mountains, and lakes more.
3.15
1. environment
2. chain
3. God
4. living
5. noise
6. God
7. healthy

3.16 Teacher check
3.17 Example: sunshine
3.18 Example: fuel
3.19 mecology
3.20 man (or you)
3.21 Teacher check
3.22
✓ pump water
feeds animals
✓ makes electricity
✓ grinds grain
✓ powers cars
3.23 where there is a lot of wind
3.24 blades
3.25 farm
3.26 power
3.27 ecology
3.28 fertilizer
3.29 melt
3.30 smog
3.31 trash
3.32 waste
3.33 windmill

SELF TEST 3

3.01
- mountains
- ✓ plants
- air
- ✓ birds
- ✓ insects

3.02
- ✓ fuel
- ✓ trash
- bank
- wind
- ✓ noise

3.03
- ride your bike
- ✓ recycle used things
- read books
- ✓ be kind to animals
- ✓ plant trees

3.04 no
3.05 yes
3.06 yes
3.07 no
3.08 yes
3.09 yes
3.010 no
3.011 no
3.012 yes
3.013 yes
3.014 no
3.015 yes
3.016 Lebanon
3.017 dirty air, water, or ground
3.018 study of environments
3.019 sunshine
3.020 you and ecology
3.021 breathe
3.022 makes plants grow
3.023 smoke and fog
3.024 repeats
3.025 around you
3.026 Example: Ecology is the study of environments.
3.027 Example: The study of ecology with me in it.

LIFEPAC TEST

Examples:

1. plants, animals, people
2. ✓ fuel
   ✓ trash
   bank
   wind
   ✓ noise
3. breathe
4. smoke
5. recycling
6. hearing
7. environments
8. no
9. yes
10. no
11. yes
12. no
13. yes
14. yes
15. yes
16. dirty air, water, or ground
17. breathe
18. you and ecology
19. repeats itself
20. sunshine
21. make plants grow
22. dominion, fowl, cattle, creeping, earth

ALTERNATE LIFEPAC TEST

1. Examples: plants, animals, birds
2. Examples: trees, fish, grass
3. live
4. environment
5. study
6. renewed
7. ecology
8. yes
9. no
10. no
11. yes
12. no
13. no
14. yes
15. yes
16. smoke and fog
17. around you
18. not living
19. world
20. study of home
21. makes plants grow
22. visitest, waterest
23. made, beautiful, time

SCIENCE 208

ALTERNATE LIFEPAC TEST

Name ____________________

Date ____________________

Each answer = 1 point

Name three treasures in your environment.

1. ____

____

____

Put a ✓ in front of three living things.

2. ________ trees

________ fish

________ fuel

________ grass

________ air

Write the word that is correct.

3. Every living thing needs water to ____________________ .

work breathe live

4. In a healthy home, there is a good ____________________ .

house pet environment

5. John Muir wanted to ____________________ the things God made.

study kill keep

6. Sunshine is ____________________ .

renewed stored lost

7. The study of ____________________ helps us find the rules of nature.

spelling math ecology

Write *yes* or *no*.

8. ________ The things around you are your environment.
9. ________ Water is not part of your environment.
10. ________ The children saw fog.
11. ________ Many old things can be used again.
12. ________ Loud music does not hurt your hearing.
13. ________ Noisy machines have more power.
14. ________ In a healthy home, the environment is kept clean.
15. ________ Fuel cannot be renewed.

Draw lines to match.

16.	smog ▸	◂	not living
17.	environment ▸	◂	study of home
18.	rocks ▸	◂	world
19.	storehouse of treasures ▸	◂	makes plants grow
20.	ecology ▸	◂	around you
21.	fertilizer ▸	◂	smoke and fog

Finish these Bible verses. Use a word from the box.

visitest	made	time	waterest	beautiful

22. "Thou ______________________ the earth,
and ______________________ it."
Psalm 65:9

23. "He hath ______________________ everything
______________________ in his own ______________________."
Ecclesiastes 3:11

SCIENCE 209

Unit 9: Changes in Our World

TEACHER NOTES

MATERIALS NEEDED FOR LIFEPAC	
Required	Suggested
• crayons • pencil • magazines • scissors • paste • plant • box • hymnbook	• Bible, King James Version

ADDITIONAL LEARNING ACTIVITIES

Section 1: People and Animals Change

1. Bring in *your* (teacher) baby picture. Ask students if they notice any change in you. Discuss growing changes (young, old, thin, tall, etc.).
2. Ask students if animals change, too. Discuss how pets grow (puppies, kittens, etc.). Ask if students know of any other unusual changes in animals.
3. Introduce the vocabulary word *lizard*. Explain the lizard's unusual color change. Introduce vocabulary word *shed*. Ask students if they can explain how and why dogs and cats shed. Finally, introduce the vocabulary word *hare*. Explain that God protects the hare by changing its color to look like what is around it.
4. Name the Baby Contest: Have each student bring in one baby picture. Arrange them on a big poster board with a number under each picture. Let students number a paper, and write their guesses as to which student belongs to which picture. The child with the most correct answers wins!
5. Have the students make their own scrapbook. Include pictures of them as a baby, one year old, two years old, and so forth. Write a few sentences for each year. Suggest that they include where they lived, favorite toys, games, friends, and so forth. Put a nice cover on it and be sure to leave extra pages for future years.
6. Assign the students to look up information on caterpillars or polliwogs. Find out how they change. Draw a picture and write a few sentences explaining the changes.

Section 2: Weather and Plants Change

1. Write the following groups of words on the board:
   1. sunglasses and swimsuits
   2. colorful falling leaves
   3. ice and snow

4. planting seeds and flowers

   Explain that each group of words reminds you of a season of the year.
   Ask students to name a season for each group of words.

5. Explain that just as the weather changes in each season, so do plants and animals. Ask students to think of ways that plants and animals change. Mention birds flying south, animals shedding fur, bears sleeping, leaves falling, and so forth.
6. Invite a florist to visit the class to show some unusual plants and explain how they change. Ask them to bring a gardenia and Joseph's coat of many colors if possible.
7. Assign the students to draw a picture and write a few sentences about their favorite season. Tell what they like best about that season.
8. Have the students soak bean seeds in wet paper towels. Check them every day for one week. Keep records of changes. When the seeds begin to show growth, plant them in a paper cup filled with soil. Water and observe the plants for two more weeks. Keep daily records of changes.
9. Assign the students to choose one kind of plant that grows near their house. Draw four pictures showing how the plant changes during each season.

Section 3: God Does Not Change

1. Discussion:

   "You have learned about all the many things in the world that change.

   Can you think of anything that does not change?

   God does not change. God's love does not change."

2. Tell the students, "Many people sing songs to praise God and thank Him for His great love for us. The songs are called *hymns*." Pass out hymnbooks or have a hymn written on the board. Sing a hymn with the students.
3. Read a Bible story about God's love to the class.
4. Have the students make up a short hymn thanking God for his everlasting love. Write it on paper or sing it to the class.

Administer the LIFEPAC Test.

The test is to be administered in one session. Give no help except with directions.
Evaluate the tests and review areas where the students have done poorly.
Review the pages and activities that stress the concepts tested.
If necessary, administer the Alternate LIFEPAC test.

ANSWER KEYS

SECTION 1

1.1 yes
1.2 yes
1.3 no
1.4 no
1.5 yes
1.6 no
1.7 Answers will vary.
1.8 grow (or change)
1.9 change (or grow)
1.10 grew
1.11 learned
1.12 learned
1.13 learned
1.14 Lynda
1.15 Lee Anne
1.16 Lynda
1.17 Mother
1.18 Lynda
1.19 Mother
1.20 a. wisdom b. stature c. favor
1.21 pillow, bowl, crow, snow, throw, blow
1.22 picnic
1.23 lizard
1.24 rock
1.25 grass
1.26 color
1.27 change
1.28 safe
1.29 true
1.30 false
1.31 true
1.32 true
1.33 false
1.34 false
1.35 true
1.36 true
1.37 true
1.38 changed as they grew.
1.39 looked fatter in winter.
1.40 shed in summer.
1.41 make you grow.
1.42 change color to keep safe.
1.43 sleep all winter.
1.44 a. bear b. hare (or cat)
1.45 hare
1.46 to look like the snow all around him
1.47 God made bears to know when to wake up.
1.48 God made the hare that way.
1.49 Teacher check

SELF TEST 1

1.01 ✓ you copy sounds you hear
you change color
✓ you learn to feed yourself
you change color of eyes
✓ your legs get stronger
✓ you grow
1.02 true
1.03 true
1.04 false
1.05 true
1.06 true
1.07 false
1.08 true
1.09 true
1.010 true
1.011 false
1.012 true
1.013 reason
1.014 grow
1.015 bigger
1.016 safe
1.017 sleep
1.018 wisdom, stature, favor

SECTION 2

2.1 changes
2.2 a. gray b. snow
2.3 a. think b. fun
2.4 a. things b. snow
2.5 Illinois
2.6 Teacher check
2.7 Illinois
2.8 a. Make a snowman
b. throw snowballs
2.9 a. water freezes
b. nose gets cold
2.10 sit by the fireplace to get warm or pop popcorn
2.11 Teacher check
2.12 a. toy b. oil
c. point d. noise
e. joy f. boy
2.13 a. every thing b. season
c. plant d. pluck
e. planted
2.14 wind
2.15 special
2.16 welcome
2.17 plants
2.18 robin
2.19 good
2.20 blossoms
2.21 The leaves turn green.
2.22 The grass turns green.
2.23 Their skin turned red and they got sick.
2.24 go on picnics and ride her bicycle
2.25 Answers will vary.
2.26 1. outside
2. bicycle
3. tired
4. warm
5. water
6. red
2.27 true
2.28 false
2.29 false
2.30 true
2.31 true
2.32 true
2.33 true
2.34 winter, colors, south, leaves
2.35 leaves
2.36 colors
2.37 winter
2.38 south
2.39 Teacher check
2.40 slowly
2.41 fast
2.42 things
2.43 light
2.44 head
2.45 red
2.46 touched
2.47 brown
2.48 trap
2.49 1. change
2.50 2. light
2.51 3. slowly
2.52 4. yellow
2.53 5. red
2.54 6. touched
2.55 7. sleep
2.56 8. leaves
2.57 Teacher check
join, choice, noise, coin, oil, coil, toy, joy, boy, boil, voice, soil, foil

SELF TEST 2

2.01 ✓ you
rock
table
✓ baby
✓ cat
✓ dog

2.02 a cat
✓ a lizard
✓ a hare
a dog

2.03 ✓ cat
lizard
✓ hare
bird

2.04 a. good b. grow
2.05 rock
2.06 fur
2.07 careful
2.08 changes
2.09 reason
2.010 slowly
2.011 sunflower
2.012 touched
2.013 smells
2.014 opens
2.015 red
2.016 true
2.017 false
2.018 false
2.019 true
2.020 false
2.021 true
2.022 true
2.023 false
2.024 true
2.025 false
2.026 true
2.027-2.030 Any order:
2.027 winter
2.028 spring
2.029 summer
2.030 fall

SECTION 3

3.1 Teacher check
3.2 true
3.3 false
3.4 true
3.5 true
3.6 true
3.7 true
3.8 Teacher check (God's Love)
3.9 point
3.10 joy
3.11 voice
3.12 boy
3.13 coin
3.14 noise
3.15 toys
3.16 foil
3.17 oil
3.18 join
3.19 Examples:
"Standing on the Promises"
"Thy Word Have I Hid in My Heart"
3.20 book
3.21 Bible
3.22 verses
3.23 you
3.24 long
3.25 hymns
3.26 Answers will vary.
3.27 Bible
3.28 Teacher check
3.29 heart
3.30 Teacher check
3.31 false
3.32 true
3.33 true
3.34 true
3.35 true

SELF TEST 3

3.01 weather
✓ God's love
plants
✓ God's Word
animals
3.02 a. bear b. cat or hare
3.03 a. lizard b. hare
3.04 cold wind
3.05 the first robin
3.06 green grass
3.07 birds fly south
3.08 in your heart
3.09 Examples: make a snowman, throw snowballs, go sledding
3.010 Any one answer: in wisdom, in stature, in favor with God and man
3.011 grow
3.012 reason
3.013 safe
3.014 season
3.015 light
3.016 sleep
3.017 sunlight
3.018 touched
3.019 circle
3.020 changes
3.021 hymns
3.022 Bible
3.023 true
3.024 false
3.025 true
3.026 false
3.027 false
3.028 true

LIFEPAC TEST

1. name
   ✓ talking
   ✓ size
   color of skin
2. lizard
   ✓ cat
   ✓ hare
   fish
3. green grass
4. cold wind, snow
5. colored leaves
6. first robin, new plants
7. wisdom, stature, or favor
8. make a snowman, throw snowballs, or go sledding
9. false
10. false
11. true
12. true
13. true
14. true
15. false
16. grow
17. four
18. change
19. leaves
20. Word
21. hymns

ALTERNATE LIFEPAC TEST

1. size, color of hair
2. lizard, hare
3. changes color
4. sleeps in winter
5. sheds fur
6. looks like a rabbit
7. Any order:
   winter, spring, summer, fall
8. true
9. true
10. false
11. false
12. true
13. false
14. true
15. trees
16. morning glory
17. gardenia
18. sun
19. circle
20. rock

SCIENCE 209

ALTERNATE LIFEPAC TEST

Name ______________________

Date ______________________

Each answer = 1 point

Put a ✓ in front of two ways you change.

1. ________ size

________ name

________ color of hair

________ color of skin

Name two animals that change color to keep safe.

2. ______________________ ______________________

Draw lines to match.

3. lizard ▸ ◂ looks like a rabbit

4. bear ▸ ◂ sheds fur

5. cat ▸ ◂ changes color

6. hare ▸ ◂ sleeps in winter

Name the seasons of the year.

7. ____________________ ____________________

____________________ ____________________

Answer *true* or *false*.

8. ________________ Some animals change color to look like the things around them.
9. ________________ The way we grow is a miracle.
10. ________________ It is always very cold in fall.
11. ________________ All plants change when they are touched.
12. ________________ God never changes.
13. ________________ Friends never change.
14. ________________ God's love is everlasting.

Write a word from the box to finish each sentence.

circle	morning glory	sun	trees	gardenia	rock

15. In the fall, leaves drop from the ______________________ .

16. The ____________________________________ is a flower that stays open in the morning.

17. If you touch the white blossoms of the ______________________________, they will turn brown and die.

18. The sunflower turns its head to follow the _____________ .

19. God's love is like a ______________________________ .

20. God's Word is called a ________________________ .

SCIENCE 210

Unit 10: Looking at Our World

TEACHER NOTES

MATERIALS NEEDED FOR LIFEPAC	
Required	
• bag or box	• drawing paper
• a variety of about 5 different objects	• pencil

ADDITIONAL LEARNING ACTIVITIES

Section 1: All About Living Things

1. Show a picture of a plant, an animal, and a person to the class.

   Discuss:

   a. How did each of these living things start?
   b. What did God provide for each of these living things so that it might be able to survive in its environment?
   c. How does each of these living things affect its environment?
   d. How might the weather and the seasons affect each of these living things?
   e. How do we take care of plants, animals, and people?

2. Review parts of living things with the students.
3. Have students draw or paint a mural of plants and animals that produce food. Encourage them to draw the food next to the plant or animal that produces it. Or, they can cut out pictures from old magazines and paste them onto the mural.
4. Have students draw a picture of one living thing and write about how it grows, how it changes, how it helps people, and how to care for it.
5. Allow students to write a report about a favorite plant or animal. Encourage them to use the library and Internet to research.
6. Make a class scrapbook about living things. Students may use old magazines to cut out pictures, draw their own, or print from online sources.

Section 2: Nonliving Things

1. Try this exercise using several different nonliving objects with the class.

   Example:

   a. Hold up a pencil. Ask if it is living. Review the differences between living and nonliving things.
   b. Ask students to describe the color, shape, size, and feel of the pencil.
   c. Ask how they could change the pencil. (They could sharpen it.)
   d. Ask how the pencil could help people.
   e. Ask how a person could take care of the pencil.

2. Play "I'm Thinking of an Object." Say, "I'm thinking of an object. Its color is ___________ . It feels ___________ . It is shaped like ___________ . Its size is ___________ ." See if the children can guess what the object is.
3. Cut out pictures of nonliving objects. Tell which objects were made by God and which were made by people.
4. As a class, go outside and take a walk around your neighborhood. Identify all nonliving objects. Discuss which ones were made by God and which ones were made by people.
5. Have the students collect nonliving objects on your walk. Allow each student to share their object with the class and describe its size, shape, color, and feel.
6. Grow some crystals as a demonstration. As a class, observe daily and ask students to describe what changes they see. Explain to students that crystals are nonliving, yet they grow!
7. Have students bring in a nonliving object to share. Students should describe:
    a. how they care for this object
    b. how this object may have changed from its original state to what it now is
    c. how this object helps them
8. Review colors with the students (primary colors, how to mix colors to get a different color, the results of mixing black or white with an existing color, and the colors of the rainbow).

Section 3: All Things Together

1. Have students close their eyes. Ask what they can hear, feel, smell, and if there is anything they can taste. Have them open their eyes and tell everything they can see. Explain that they have just described their environment. Review the definition of environment.
2. Ask the class to imagine that they are in a far away strange land. Ask them to write or tell about their strange environment. Students may draw pictures to accompany their stories.
3. Divide the class into groups of 3 to 4 students. Have each group list at least two good health habits. Combine all the ideas into a class chart which can be displayed. (The chart may be illustrated by hand, or students may cut out and paste pictures onto it.)
4. Students may draw a picture to show how they take care of their environment.
5. Have students write a report telling why it is important to practice good health habits.
6. Have the students discuss and draw pictures about how living and nonliving things change when the seasons change.
7. Go on a nature walk. Identify all living and nonliving things. Discuss how the things you see are affected by the weather and the seasons.
8. Divide the class into several teams, and go on a scavenger hunt. Make sure to include both living and nonliving things on the hunt lists.
9. Assign the students to write a short story or tell about their environment. Encourage students to describe living and nonliving things as well as what they can see, hear, smell, taste, and touch.
10. Watch a local weather report together and discuss how your environment will be affected by today's weather.

11. Discuss with the students how they can care for their environment. Choose a clean-up project to do together as a class.
12. Create a class terrarium. Include both living and nonliving objects in your project.

Administer the LIFEPAC Test.

The test is to be administered in one session. Give no help except with directions.
Evaluate the tests and review areas where the students have done poorly.
Review the pages and activities that stress the concepts tested.
If necessary, administer the Alternate LIFEPAC test.

ANSWER KEYS

SECTION 1

1.1 adult
1.2 plant
1.3 chicken
1.4 dog
1.5 horse
1.6 Any order:
a. light b. food
c. water d. air
1.7 Teacher check
1.8 Examples:
run — ran or Ron
hat — hit, hut, or hot
pet — put, pit, pat, or pot
dog — dig or dug
leg — log, lug, or lag
1.9 cup
1.10 leg
1.11 red
1.12 fell
1.13 sick
1.14 Any order:
a. spring b. summer
c. fall d. winter
1.15 The leaves change color and fall off the tree.
1.16 Yes, The black circle got bigger in the dark and became smaller after the light was turned on.
1.17 a. lost a tooth
b. washed or is clean
1.18 fat
1.19 shed
1.20 sleep
1.21 Answers will vary:
a. place to live for animals
b. food for animals
c. water for plants and animals
d. sunlight for plants
e. cleaning and washing animals, cutting grass, pulling weeds
1.22

baby	**ice**	**peas**
maple	side	tree
rain	ride	seed
take	bite	bean
pony	**use**	
grow	cute	
open	cube	
old	mule	

1.23

P	apple	A	meat
A	eggs	A	milk
P	bean	P	orange
P	peas	P	corn

1.24 <u>ship</u>, <u>shed</u>, <u>shoe</u>, <u>show</u>
<u>thin</u>, <u>three</u>, <u>think</u>
<u>chin</u>, <u>chick</u>, <u>chain</u>, <u>chair</u>

SELF TEST 1

1.01 plants, animals, people
1.02 air, sun, food, water
1.03 spring, summer, fall, winter
1.04 making food, doing work, giving cotton or yarn for clothes
1.05 X
1.06 circle
1.07 X
1.08 circle
1.09 X
1.010 circle
1.011 X
1.012 X
1.013 food
1.014 yarn
1.015 sheds
1.016 grows
1.017 washing
1.018 sheep
1.019 chick
1.020 thin
1.021 e circle
1.022 i
1.023 a circle
1.024 o
1.025 u
1.026 e circle
1.027 o
1.028 o circle
1.029 Yes, They give us food. They do work. They give us cotton and yarn for clothing.

SECTION 2

	OBJECTS	BIG	LITTLE
2.1	P		✓
2.2	G	✓	
2.3	G	✓	
2.4	P		✓
2.5	G	✓	
2.6	P	✓	

2.7 a. Any order: red, yellow, blue
b. Any order: red, orange, green, blue, indigo, violet
c. 1. orange
2. purple
3. green
4. light green
5. dark red
6. gray

2.8 Teacher check
2.9 triangle
2.10 circle
2.11 irregular
2.12 rectangle
2.13 size, shape, color, feel
2.14

fork	**car**
corn	star
born	arm
horse	barn

2.15 horse, barn
2.16 wood, nails
2.17 saw, hammer
2.18 He made them into a table.
2.19 3 – The Jones family has a new table.
1 – Mr. Jones went to the store.
2 – Mr. Jones pounded nails.
2.20 melted
2.21 got bigger
2.22 Teacher check
2.23 Drawings will vary.
2.24 s
2.25 s
2.26 g
2.27 g
2.28 k
2.29 j
2.30 s
2.31 k
2.32 Teacher check
2.33 Teacher check
2.34 bird, chirp
2.35 hammer
2.36 fur
2.37 see
2.38 hear
2.39 smell
2.40 taste
2.41 touch
2.42 Drawings will vary.
2.43 Thinking about God's Gifts
2.44 Answers will vary.
2.45

star	**corn**
car	fork
arm	horse
barn	store

hammer, fur, bird
paper
chirp
purple

SELF TEST 2

2.01 no
2.02 yes
2.03 yes
2.04 yes
2.05 no
2.06 yes
2.07 yes
2.08 no
2.09 yes
2.010 yes
2.011 Any order: taste, smell, hear, see, touch
2.012 black
2.013 small
2.014 circle
2.015 hard
2.016 The sun can melt the snowman.
2.017 You can write words and draw pictures with a pencil.
2.018 j
2.019 k
2.020 s
2.021 car
2.022 paper
2.023 fork
2.024 bird
2.025 purple

SECTION 3

3.1 in a cabin in the woods
3.2-3.4 Answers will vary.
3.5 summer
3.6 Drawings will vary.
3.7 Answers will vary.
3.8 Answers will vary.
3.9 Answers will vary.
3.10 yellow
3.11 grow
3.12 blow
3.13 flower
3.14 town
3.15 summer fall
winter spring
3.16 Examples:
a. boy's clothes
b. weather or tree, flowers, lake
3.17 house
3.18 flower
3.19 mountain
3.20 cow
3.21 weather
3.22 cloudy
3.23 rains
3.24 thunder
3.25 snows
3.26 Teacher check
3.27 pollution
3.28 noise
3.29 odors
3.30 Examples:
a. washed windows
b. picked up papers or took out trash
3.31 He fixed the car.
3.32 Examples:
pick up cans and bottles
throw trash into trash cans
3.33 coin
3.34 Joy
3.35 boy
3.36 boil
3.37 toy
3.38 noise
3.39 Teacher check
3.40 ✓ Brush your teeth after meals.
Eat lots of candy.
✓ Eat good food.
Stay up late at night.
✓ Play outside.
✓ Wash your hands before meals.

SELF TEST 3

3.01 Teacher check
3.02 environment
3.03 cabin
3.04 odor
3.05 thunder
3.06 brush
3.07 food
3.08 weather
3.09 pollution
3.010 Any order: color, size, shape, feel
3.011 noise
3.012 cow
3.013 cloud
3.014 boy
3.015 house
3.016 town
3.017 toys
3.018 snow
3.019 grow
3.020 Examples: to feel good, to grow, to use your mind to learn

LIFEPAC TEST

1. red, yellow, blue
2. adult
3. food
4. brush
5. sheep
6. seed
7. warm
8. drop
9. fat
10. car
11. mind
12. color
13. size
14. feel
15. shape
16. environment
17. has no leaves
18. washing windows
19. swim
20. picking up trash
21. smell
22. hear
23. see
24. taste
25. feel
26. ✓ your mind works better
    you can go to the store
    ✓ you can feel good
    ✓ you can have more fun enjoying God's world
    you must stay home from school
27. Examples: seeds, animals, pets
28. Examples: read books, ask teacher

ALTERNATE LIFEPAC TEST

1. Any order:
   a. light
   b. water
   c. food
   d. air
2. Any order:
   a. spring
   b. summer
   c. fall (autumn)
   d. winter
3. Any order:
   plants, animals, people
4. Any order:
   a. size
   b. shape
   c. color
   d. feel
5. Nonliving things
6. God
7. People
8. primary
9. circle
10. Irregular
11. environment
12. sight, hearing, taste, touch, smell
13. how hot or cold it gets where you live
14. made in the image and likeness of God
15. the place where you live
16. something that makes the environment dirty
17. hair that grows on sheep
18. to lose hair or fur
19. a smell

SCIENCE 210

ALTERNATE LIFEPAC TEST

Name ______________________

Date ______________________

Each answer = 1 point

Answer these questions.

1. In order to grow, all living things need what four things?

a. ______________________ b. ______________________

c. ______________________ d. ______________________

2. Name the four seasons.

a. ______________________ b. ______________________

c. ______________________ d. ______________________

3. Three kinds of living things are:

a. ______________ b. ______________ c. ______________

4. Name four ways to describe objects.

a. ______________________ b. ______________________

c. ______________________ d. ______________________

Circle the correct answer.

5. ______________________ do not drink water or eat food.

   Living things   Plants   Nonliving things

6. ______________________ made objects like the sun, moon, oceans, land, and sky.

   God   People   Companies

7. ______________________ make objects like houses, cars, and toys.

   Animals   God   People

8. Red, yellow, and blue are called ______________________ colors.

   pretty   primary   nice

9. A ______________________ is a round and flat shape.

   circle   triangle   square

10. ______________________ is a word used for shapes that are not like any shape you can name.

    Regular   Odd   Irregular

11. All of the living and nonliving things that are around you make up your ______________________ .

    climate   likeness   environment

Draw lines to match.

12.	five senses ▸	◂ the place where you live
13.	climate ▸	◂ sight, hearing, taste, touch, smell
14.	you ▸	◂ to lose hair or fur
15.	environment ▸	◂ made in the image and likeness of God
16.	pollution ▸	◂ hair that grows on sheep
17.	wool ▸	◂ a smell
18.	shed ▸	◂ how hot or cold it gets where you live
19.	odor ▸	◂ something that makes the environment dirty